Emerging Leadership: Black Tech Voices Redefining Executive Excellence

Emerging Leadership:
Black Tech Voices Redefining Executive Excellence

ITSMF INFORMATION TECHNOLOGY SENIOR MANAGEMENT FORUM

INFORMATION TECHNOLOGY SENIOR MANAGEMENT FORUM PUBLISHING

itsmf INFORMATION TECHNOLOGY SENIOR MANAGEMENT FORUM

Published by:

Information Technology Senior Management Forum

Itsmfleaders.org

@itsmfleaders

First Edition: October 2023

Book Designer: Dawn Jordan Jones | 29Eleven Media

ISBN 978-1-7365068-5-1 (paperback)
ISBN 978-1-7365068-4-4 (ebook)

Table of Contents

Foreword: *Message from the Editors*

Welcome to the 2023 edition of the ITSMF Executive Academy capstone book on leadership. The individual essays are written by senior Black technology leaders for Black technology leaders who are forging paths of excellence in a world where challenges and opportunities intertwine like never before. As we gather the experiences and insights of exceptional Black leaders in the tech industry, we set sail on a journey that celebrates resilience, brilliance, and unwavering determination to shape the future.

In the realm of technology, the landscape is ever evolving, filled with twists and turns. Our ITSMF Executive Academy leaders today stand at the intersection of innovation and change, with the unique perspective and experiences that come with being a Black leader in this space.

This book is more than just a collection of personal stories, career histories, or a compilation of leadership theories. It is a vibrant tapestry of voices, each thread telling a tale of inspiration, triumph, and the occasional stumble. Through these pages, you will hear from pioneers who have charted new territories, faced obstacles head-on, and, against varying odds, shattered ceilings that were meant to remain unbroken.

We recognize that the journey of a Black technology leader is a path less traveled, filled with challenges that our non-Black counterparts may never fully comprehend. But it is precisely those experiences that have shaped our Executive Academy members into the remarkable leaders they are today. Their ability to navigate through adversity and still bring brilliance to their work is nothing short of remarkable.

Within these chapters, you will find real-world advice, anecdotes, and battle-tested strategies to help you harness your leadership potential and confront the wide range challenges you may face along the way—from imposter syndrome to systemic biases, from diversity and inclusion to fostering innovation, you will gain valuable insights that will elevate your leadership to new heights.

Our intention here is to amplify the voices of Black technology leaders, spotlighting their extraordinary contributions which are often underrepresented and overlooked. We believe that the stories and wisdom shared here will not only inspire the next generation of Black leaders but also spark meaningful conversations about diversity, equity, and inclusion, within the tech industry and beyond.

To the 2023 Executive Academy Cohort: Thank you for joining us on this transformative journey. Your leadership is a beacon of hope, shining brightly and guiding others through the storms that may come their way. Remember, together, we are leading the way.

Robert Scott
Vice President & Dean, ITSMF

Dr. James Rodgers
Director, ITSMF Executive Academy

I

How Leaders are Shaped:

Lessons Learned, Hard and Otherwise

What Mom Knew About Leadership

by Lisa Godfrey

As a child, I was always curious about the world around me. I loved asking questions, exploring new places, trying new things. It was my mother who saw my potential as a leader and worked tirelessly to help me develop the traits I needed to make a positive impact on the world. She encouraged me to take on leadership roles in school, Girl Scouts, church, and other activities such as sports and volunteer work, so I could learn valuable skills like teamwork, communication, goal-setting, and problem-solving. She also made sure I had access to books, movies, and other resources to inspire and motivate me. I was empowered to embrace my unique qualities, interests, and personality, and as a result I was able to develop organically without conforming to the expectations of others or trying to portray someone I was not. I learned to be accountable for my actions and take responsibility for my own growth, development, and success.

My mother's influence went far beyond just providing opportunities for me to learn and grow. She offered guidance, support, and encouragement, listened patiently as I shared my hopes, dreams, and fears, and helped me navigate adolescence and young adulthood with grace and resilience.

As I grew older and began to take on more responsibilities and challenges, my mother continued to be my biggest cheerleader. She celebrated my successes but also helped me to learn from my failures and mistakes. She taught me the importance of resilience, perseverance, and humility, inspiring me to keep pushing myself to be better, even when things got difficult.

I realize now that my mother was not just raising me to be a leader. She was teaching me how to be a compassionate, thoughtful, and empathetic human being who cares deeply about the world and the people in it. For that, I am forever grateful.

These youthful experiences led me to a fulfilling and purposeful adult life, a life of leadership. Now, I strive to inspire others to become thoughtful, effective leaders. Throughout my 25-year career in the technology industry, I have observed various leadership styles; some have been inspirational, and others have made me want to run far and fast. While I have differing perspectives about each leadership style, I adopted bits and pieces to help shape and transform me as an influential leader.

***

Executive leadership is a critical component of any organization, involving guiding and directing a team toward achieving its goals, creating a clear, concise vision and strategy, and communicating it effectively to employees. Executive leaders play a crucial role in driving success and achieving organizational goals by evaluating complex situations, identifying priorities, and formulating strategies to address them. These leaders have a deep understanding of their industry and the broader market, enabling them to make informed decisions that position their organizations for growth and success. By developing and executing wise plans, executive leaders can drive innovation, enhance performance, and create value for stakeholders. Their ability to inspire and motivate teams, while staying focused on achieving outcomes, makes them invaluable assets. An effective executive leader inspires and motivates others to work toward a common goal, demonstrates professionalism and integrity, is committed to continuous education and self-development, and can quickly adapt to changing circumstances.

Following my mother's example, I'm passionate about the need for executive leaders to make a commitment to continuous learning and improvement. Leaders must be open to feedback and willing to learn from others, including employees, customers, and industry experts. They also need to be willing to invest in their personal development, through training, coaching, and networking. This helps to ensure that the leader is up to date on industry trends and best practices and can make informed decisions.

***

My mother also taught me to ask for help. As an executive leader, it can be easy to fall into the trap of thinking you must have all the answers. However, asking for help when needed can be a sign of strength and effective leadership. It shows a willingness to admit when you don't know something or when you need support, which can inspire trust and respect among your team and organization. Asking for help can also lead to better outcomes. When you seek input and ideas from others, you benefit from diverse perspectives and expertise, which can lead to better decision-making and problem-solving.

When asking for help, it's important to be specific about what you need and to approach the situation with humility and an open mind. Acknowledge the contributions of others and be willing to give credit where credit is due, which encourages trust in a team.

Trust goes hand in hand with transparency in executive leadership, the degree to which leaders communicate openly and honestly with their stakeholders, including employees, shareholders, customers, and the public. Transparent leaders prioritize clear communication, actively seek feedback, and make decisions based on facts and data. They create a culture of trust and accountability by being open about their thought processes, goals, and challenges. Transparent leadership emphasizes inclusion and collaboration among stakeholders, leading to better decision-making and improved organizational performance. This promotes ethical behavior and reduces the risk of corruption or unethical practices.

Transparency connects to the issues of ethics and integrity and the need for leaders to hold themselves and their teams to high standards of conduct—something my mother modeled for me. This includes being honest and open with stakeholders, avoiding conflicts of interest, and protecting the organization's assets and reputation. The leader must be willing to speak up if they see unethical behavior within the organization and take steps to address it appropriately. Leaders who act with integrity and hold themselves and others accountable for their actions will earn the trust and respect of their constituents.

***

One last point, which T.D. Jakes said best: "If you have achieved any level of success, then pour it into someone else. Success is not success without a successor." This emphasizes another valuable lesson I learned from my mother—the importance of empowering and uplifting others.

Lisa M. Godfrey

Lisa Godfrey is a Technology Manager for Bank of America with over 25 years of experience leading high performing teams. Lisa's team is responsible for managing testing efforts, developing testing methodologies, and creating standards and procedures for quality assurance testing methodologies for the Chief Administrative Office and Global Human Resources technology organizations within Employee Experience Technology.

Do Leaders Really ‘Do Nothing’?

by Kay Malcolm

The following is a fictionalized story that captures leadership lessons from the ITSMF Executive Academy experience. Similarity to real or fictional characters or organizations are unintended and coincidental.

In the fast-paced, achievement-based world of corporate America, perhaps no two words strike fear more than: Do Nothing. Yet, if we believe current teachings about executive leadership, doing ‘nothing” is at the core of the role—and the one thing that most executive leaders get wrong.

<u>One Story: Journey to the Top. Not so Easy.</u>

When I get to Otercel, I am going to be a leader. Astrid was always a go-getter. He graduated top of his class in high school and college, a bright light who expected the future to continue in that pattern. When he joined the company, he was inspired by the Black president, the way he moved, the way he carried himself, his reputation in the industry. *I want to be him in 10 years*. And so, Astrid spent the next 10 years struggling toward that end but was no closer to leadership than the day he confidently made the claim. He was frustrated, felt mal-aligned, the victim of constant discrimination by his white male peers who seemed to rise easily along with their white fraternity brothers and networks he was unable to penetrate. “The system is rigged,” he decided.

Then Astrid read an article that changed his perspective. The article advised that successful executives distinguish themselves not only by good performance, but also by having mentors and sponsors (often white men). Further, that successful leaders make themselves available for larger roles, and never complain or got derailed by discrimination.

Inspired, Astrid dove in with renewed perspective, vowing to play the game in front of him, not try to change the game. He began with learning more about his company. Leaders must understand four essential components of their company. Otercel's success came down to cash, profit, assets, and people. Suddenly, Astrid saw clearly how critical it was for everyone in the company to know how their work contributed to the company's success via those four components. He began diving into Otercel's financial statements, learned how the company made its money. Then he showed that he was open to new projects, and stopped getting distracted by non-essential initiatives, keeping his eyes on the prize—and had an epiphany. Otercel's goal was no different than the lemonade stand he and his sister had operated on hot Florida days—to make more money than it spent. When communicating with his leadership, Astrid's responses addressed how they could better position Otercel in the marketplace in order to win.

Almost immediately, Astrid's boss was asking him to join meetings he had once begged to attend, and he left one meeting with his first leadership position, to drive the onboarding of a recent acquisition, a smaller outfit that Otercel had recently purchased.

***

"I can't believe that three years ago, I almost gave up my dream of leadership," Astrid said, leaning back and reflecting with his wife. "We made it babe, we've arrived." Elle looked at him, laughed and said, "Don't get comfortable here, you remember what happened to Marsha." As much as Astrid hated to admit it, his wife was right. His college electrical engineering study partner, Marsha, was no longer with the major software developer she'd joined. She had become complacent as she rose, holding to the notion that she was the only one who could fix things; she didn't trust her people, and that became her undoing.

Astrid set out to identify his own blind spots and began to ask for feedback from colleagues. What he heard was that he was more of a manager than a leader. Yes, his title *was* "manager," but leadership was what he was after. So, he went about doing his homework, first reading *What Got You Here Won't Get You There* by Marshall Goldsmith—and resolved to incorporate some of the recommended precepts:

- Move from being task oriented to intuitive and visionary.
- Reduce focus on maintaining order and stability and focus more on innovation and change, motivating the team to think outside the box.
- Be more comfortable taking risks.
- Drive innovation and growth in decision making.
- Promote lieutenants into manager roles so they can play complementary roles and help maintain order and stability.

After three months, Astrid's boss, Jenny, peeked her head into his office. "Astrid, I'm on my way to a meeting with senior leadership, will you join me? We are planning some changes with the business, and they want to hear your perspective."

During the meeting, the Executive VP boomed from his end of the table, "Astrid, your work with the acquisition of Rivetink was well done. You have proven yourself ready for more. Are you up to the challenge?"

"Y-yes, yes Paul, I'm ready, let's do this." Astrid swallowed hard. Paul was challenging Astrid to level up their legacy business, where many executives had been working since Astrid was in elementary school. To level up, change was key. Two of his peers had tried and failed miserably and were now on "permanent vacation."

Astrid walked slowly back to his office, his mind churning, fear gripping every step. Sitting down at his desk, he looked at his bookshelf, laughing to himself. *I thought if I read every book on leadership, I would be ready for this moment. But I'm not—yet.* His eyes fell on a book whose cover was ripped to reveal the word *change*. He jumped up, that's it. *I'm* the solution! *Change* is the solution! Astrid went to his whiteboard and began creating a vision for this change. He listed: reaching out to his network, building a coalition of supporters, learning how to get buy-in from the stakeholders. His plan was to be transparent, open, and focused on the success of others. The journey wasn't easy, and when they hit roadblocks, he adjusted the approach and ensured continued change efforts by staying on track, knowing that this transformation was not a one-time event but a process.

***

"My journey to this place in my career can be summed up with five key points. Harvey Coleman said it best. I would like to share it with you," Astrid said. He was back in St. Lucia, Florida, speaking at his high school. Elle was at the side of the stage grinning, his daughter hiding behind her legs overwhelmed at the audience of students, teachers, and at least half the parents, there to see how this hometown boy had gone from playing soccer barefoot to one of the most sought-after executives in the US.

"Sixteen years ago, I made a vow to myself that I would be a leader. Here are the key things I did, motivated by Mr. Coleman's advice."

- Be self-aware. Know your strengths and weaknesses. Develop weaknesses, leverage strengths.
- Relationships are everything. Build them. Get several mentors. Find sponsors who don't look like you; they are key for visibility, credibility, and growth.
- Take risks. Each time I stepped out of my comfort zone, new opportunities and experiences followed. They weren't all easy, I failed fast, used it as a learning opportunity, and kept at it.
- Communication is critical. If you are to build relationships that influence and help you achieve

goals, then your communication, both verbal and non-verbal, must be well developed.

- Your mindset matters. Be a lifelong learner. Learning for me didn't end when I left these walls. I sought new experiences and learned how to be resilient and successful in the face of change.

***

Astrid paced in his office. "I'm torn, I don't know what to do," he said to Paul. The executive council that he was a part of was struggling to respond to a horrible police incident that had the entire nation on edge.

"Astrid, we need to tackle this diversity thing, you are the best person to lead it. We are going to create a new position, Chief Diversity Officer; you will report up into HR. Listen, I know it's not engineering but I'm in a tough spot and I know I can count on you. Give me your answer by end of day," Paul said as he stood and walked out.

After a day of pacing that Astrid was sure had worn down the carpet in his office, he walked into Paul's office. "Paul, I have your answer."

Paul looked up from his laptop, "Yes? This D&I thing, the pressure from women and people of color. We do need to get in front of it so that we are on the right side."

"Agreed Paul," Astrid said, adding, "but I have a slightly different solution that will work. We are focused on the wrong thing. We are using the word *inclusion* to be deliberately *exclusive*. I want us to tie our DEI efforts to real business outcomes and include everyone in the conversation, not just Black people and women but also our veterans and our aging white men who feel left out of the conversation. Let's make this a broader strategy, get over the distraction of differences and work on this as a team. I am an engineer turned executive; I want to stay on as CTO."

As Paul's eyes narrowed, Astrid braced for his response, and finally Paul extended his hand. "Astrid, yes. Business outcomes. We are all diverse, let's all work on it so no one feels left out." Astrid breathed out a sigh of relief, winked at Paul's white male veteran assistant as he walked out the door. He was including everybody to ensure success for everyone.

At home, Elle asked her husband, "How did you not faint, saying no to Paul?"

"I expressed my perspective and we worked together to find a mutually beneficial outcome. If I'm being honest, I half expected him to fire my butt. I was so angry that he asked me, the CTO, *and* the only other person of color on the council to be the Chief Diversity Officer. But I decided I would avoid making judgements on Paul and went into *crucial conversation* style."

"Is that the book you've been reading for the last week?" Elle asked.

"Yep! *Crucial Conversations: Tools for Talking When Stakes are High*. I loved what Kerry Paterson and the other authors of that book had to say." He took a sip of coffee, and continued, "I have said yes to every challenge Paul has given me, but this one was not for me. I needed a playbook to be calm and respectful, even though my emotions were screaming. I pulled the book off the shelf, read a section, said a prayer, and walked into that office."

Astrid exemplifies the need for leaders to stay curious learners. Astrid got it right. He embarked on a journey of discovery, networked, built relationships, took risks, focused on innovation, inclusiveness, sought feedback, all to position himself for success, ultimately making a meaningful impact on his organization and industry.

Kay Malcolm

Kay Malcolm is a Senior Director, Database Product Management at Oracle. She has held a number of positions in large software companies and Big 5 consulting firms. Her most recent roles have been focused on the Oracle technology stack where she has led global teams to effectively execute their charter.

Valuing People, Welcoming Discomfort

by Dana McKinney

Over the course of my 21-year career as an information technology professional in corporate America, I've worked tirelessly to advance my professional knowledge and technical know-how and refine my interpersonal skills in pursuit of becoming a more well-rounded executive leader. Obtaining the support required and successfully executing has at times felt very much like spiritual, intellectual, and emotional warfare. While I have undoubtedly earned my fair share of scars as I've racked up wins and losses, I've also banked a tremendous amount of value from each experience. I hope that sharing this body of knowledge helps by providing you with practical tools to deploy during your own executive leadership journey.

I've encountered many corporate professionals who proudly boasted of honing their leadership skills and principles after years in executive leadership and benefiting from one-to-one coaching and mentoring from a well-respected, well-seasoned C-suite leader. Well, this was not the case for me. My journey towards understanding executive leadership was very different.

Leadership Principle Number one: Value People

Back in 2012, I was wrapping up what had by all accounts been a very successful two-year work assignment in Japan. I'd been given the opportunity to lead our Japan affiliate and manufacturing site through the transformation out of our legacy enterprise resource planning (ERP) system and into our global instance of SAP. After successfully completing the assignment, I was promoted and offered a rather unique role supporting one of our enterprise IT organizations based out of our US corporate office. I was excited because in my mind

I had *arrived*. I was being recognized and getting more scope, receiving a substantial pay increase packaged with the promotion, and most importantly I was excited to now be reporting directly to the executive officer responsible for the organization.

What transpired over the next two years reporting to that leader helped set the foundation for what I believe is one of the most important leadership principles I've acquired to date. The principle is simple: people who feel included and that their contribution to the team is valued, will perform at a higher level than those who are not! I believe work environments that foster this inclusive approach and recognize an individual's sustained contribution, stimulate creativity and organically foster a high-performing team culture. I'm sorry to say, I did not learn this from my supervisor, who did not practice this principle. In fact, I'm not sure he knew it existed.

I spent the first 12 months in that role under constant assault and feeling I'd made the biggest mistake of my life. I walked into the role eager to contribute, ready to collaborate, and excited to be part of an amazing team, but left after nearly two years feeling defeated, rejected, and completely alone. I'd gotten a first-hand lesson in what it meant to be a good leader, not by observing good leadership in practice, but by being up close and personal to what I considered a poor leader. While one of my most challenging and disengaging work experiences, in the end this proved a very fruitful experience as it accelerated my own professional leadership development journey and created a clear leadership contrast that I have since built from.

***

For as long as I can remember, my parents preached to my siblings and me the importance of finding value from every experience, even if those experiences severely challenge you. I grew up in a home where my dad, an officer with the Los Angeles Police Department, returned to college in his early thirties and became a chiropractor, all while supporting a young wife and three young sons. Only now, after raising four children, do I understand and more importantly, appreciate that sacrifice. My father would tell his children (quoting Les Brown), "Be willing to do the things that others won't do, so that you can one day have the things that others don't have." Usually, he'd say this just before making my brothers and me go outside to clean the yard. I must admit, I did not always understand, but now I do, and wow is it prophetic!

The treasure trove of advice gifted by my father includes my second leadership principle.

Get Comfortable Being Uncomfortable

Over the course of my career, I've had the pleasure to serve in varied IT roles, delivering value to many different customer groups. I can tell you without a shadow of a doubt that, as Dad advised, I've valued each of those unique experiences and the relationships built along the way. I've spent the last ten years working in our manufacturing organization, helping to improve the lives of patients around the world, and taking advantage of opportunities to invest my time and experience in developing the next generation of IT leaders.

After a few years in rotations at manufacturing sites, leading IT teams in the US, I was promoted into the role of Manufacturing IT leader, accountable for overseeing IT operations at our Asia Pacific manufacturing sites. My family and I packed up and relocated to China where we set up our home base for the next three years. Talk about deliberately moving away from your comfort zone! I can still recall those first few months, feeling overwhelmed, off balance, and at times completely alone. The decision to relocate directly touched my wife and my children as well as me. We'd made the decision to remove ourselves from the comfort of our home, our community, and the network of family and friends who had provided love and support over the years. While operating outside of our comfort zone, we were determined not to focus on the things that made us feel lonely and distressed, but instead on all the opportunities that amazing international experience was providing. As we embraced learning the language, trying the local cuisine, being more deliberate about discovering the history and culture, our comfort level comfort increased. That growth also began to show up in my relationships with my peers and my team. As those relationships improved, my team's performance improved.

The more we leaned into this new experience, the more comfortable we became at being *uncomfortable*. We began to enjoy the experience more fully and those same successes showed up in the kids' school grades, and work successes for my wife. I'm proud that in 2021, my wife and I repatriated back to the US with two teenage daughters, both bilingual, well-traveled, and well-versed in world culture.

***

Early in my first site leadership role, I was asked to take on additional responsibility in an area of manufacturing of which I had very little knowledge. The increased scope was exciting, but there were times when I felt a near overwhelming sense of fear because, knowing I still needed to deliver. It was this experience that reinforced the importance of my final leadership principle.

<u>Make sure you are getting the right people on the right bus at the right time.</u>

As I've developed leadership capability over the years and been awarded with opportunities to experience new assignments in the US and abroad, there remain two consistent themes. First, I'm always the new guy. Second, I almost certainly don't know nearly as much as those senior, more established members of the team.

What I've learned about being a good leader is that as you grow and embark on new opportunities, your success is in large part dependent on your ability to identify the local influencers and experts, develop meaningful and mutually beneficial relationships with those people, and then leverage those relationship to deliver outcomes that matter to everyone. Ask yourself: *Do you know what technical skills and work experiences your team members aspire to achieve? Do you know their career dreams and goals? More importantly, can you advance the mission of the enterprise by helping your team members turn their dreams into reality?* By connecting the dots on these three areas, you create a win/win/win environment where you, your team members, and the enterprise all benefit.

After twenty-one years of service, here's what I know about the various initiatives that did not go as I'd planned. Most of those deviations were a direct biproduct of my neglect in properly investing in the right relationships, with the right people, at that right time. But when I've had success, I attribute it to a genuine interest in wanting to learn, an insatiable desire to collaborate and win *together*, and a willingness to overcome my own social awkwardness in pursuit of finding commonality with my colleagues.

Proactively building those mutually beneficial connections requires practice and almost always comes pre-packaged with a free cup or two of humility. But it's well worth the effort. Like with all muscles, the more you practice, the stronger and more natural connecting will feel, so don't let a lack of grace deter you. Building these connections will improve engagement and increase your collective chances of winning. The message here is simple: your success is in part dependent on your ability to develop mutually beneficial relationships with those around you. Ensuring you are getting the right people, on the right bus, at the right time is a critical factor in determining the success of every project, of every organization, and of every executive leader.

***

My three hard-won foundational principles of leadership have served me well. It's my hope they will serve you well as you pursue your own executive leadership aspirations. Strive to build meaningful human to human connections founded in a genuine curiosity and respect for people, with attention toward embracing daily challenges head on—especially those that make you uncomfortable—and surround yourself with the right people at the right time. Everyone wins!

Dana Garland McKinney

Dana McKinney is Senior Director, Manufacturing and Quality, Tech@Lilly. He has 21 years of rich IT experience that has enabled him to deliver sustainable value across various enterprise and manufacturing IT organizations in the US and abroad.

My Own Way to L.E.A.D.

by Ludell McRae

As a young girl, my dream was to become a secretary. I admired the professional way secretaries dressed and that they worked in an office. My parents are first generation immigrants who worked in factories and the service industry. One day I told a friend that I wanted to be a secretary, and she asked me why I didn't aspire to be the boss and not the secretary. I was perplexed. She said you can still dress nicely, plus have the power to tell others what to do. This conversation laid the foundation for me to embrace the notion of leadership. TV shows such as *Who's the Boss, The Jeffersons*, and *The Cosby Show* started popping up, with women and people of color starring in successful, non-traditional roles, debunking the thinking that leadership is a rite of passage and reserved for a certain group.

I've often wondered, if the management of human capital is such a critical element for a company's competitive advantage, why is it that the skills to motivate, inspire, and guide are recognized in only a small group of people? People are compensated well for their tactical achievements rather than the intangible ones—like leadership skills, which are often considered soft skills, nice to have but not essential.

Companies spend copious amounts of money and time sending employees to annual leadership training, often with few results. According to TrainingIndustry.com, leadership development is a $366 billion global industry. Yet most programs don't work. How can companies avoid over-investing in leadership programs which do not work? I too, have spent countless hours attending leadership courses and reading books about leadership. My conclusion? Leadership is not a skill developed in a classroom. The current thinking of what leadership means—the ability to sell an idea, a concept or vision in such a way that people are willing to follow—depends on breaking down barriers, building relationships, and being vulnerable; these are basic human traits yet take years to refine and master.

My empathetic nature and desire to help others succeed drives me to build meaningful connections, based on understanding what motivates people and decoding their communication style. I enjoy taking note of people's hobbies, children's names, and vacations so I can strike up conversations and show real personal interest. One-to-ones are moments to get to know people. My extroverted personality means this comes naturally for me. For some, it requires more of an effort, but the reward is well worth it. People enjoy working with people they know. People they trust. People they like. This revelation has pushed me to fight my natural tendencies of keeping work at work and my personal life private. By embracing the techniques of storytelling, I have learned how to control my narrative while still safeguarding my privacy.

My journey and experience with the ITSMF Executive Academy has been one of discovery. Unlike many other programs, this course breaks down barriers, exposes blind spots, guiding you toward the skills required to be an effective leader. An early lesson is to understand who you are. It is mind-boggling how a seemingly cliché personality questionnaire helped me grasp the foundation for effective leadership: knowing yourself first. From there, you can accept other people's flaws and limitations while enhancing the traits which make them rock stars.

The ITSMF Executive Academy reignited my desire to not settle for being a *good enough* leader or succumbing to what is referred to as an "absentee leader"—someone in a leadership role who is psychologically absent from the role and the people they lead. This person may enjoy the privileges and rewards of their leadership role but avoid meaningful involvement with their teams.

I once took on a role once in which I was told that the team pretty much runs itself and my role was to "keep the lights on and doors open." My interaction with the team revealed a different story. They appreciated my high touch nature and valued being led.

Sheryl Sandberg describes leadership as "…making others better as a result of your presence and making sure that impact lasts in your absence." I'm still working toward mastering the art of leading with this purpose.

There is a belief that there is a never-ending supply of talent willing to work at any organization for the right price. Yet according to OC Tanner research, 79 percent of employees who quit their jobs cite lack of appreciation as a primary reason. Large bonuses and payouts can never replace the simple act of saying "Thank you" which doesn't cost anything and requires very little effort while leaving a lasting impression. People seek and yearn to be recognized and feel appreciated. Even Spock, on *Star Trek* once stated, "The needs of the many must outweigh the needs of the few."

Nothing is ever about the leader, but about the people they serve.

***

I have adopted and modified something presented during ITSMF seminars, the L.E.A.D. framework, to better reflect my leadership style -- **L**isten, **E**levate, **A**ccept, **D**evelop.

- **Listen**. Leaders must listen attentively to their people to better understand their needs and areas for growth. An environment in which people feel safe to express concerns and feedback is essential; leveraging the art of storytelling builds connections and relationships.
- **Elevate.** Elevating others leads to empowerment, encouraging people to develop the growth mindset by utilizing their creative thinking and innovative inclinations to reach their full potential. Being empathetic lets people know that you care. In the inspirational words of Maya Angelou: "I've learned that people will forget what you said, people will forget what you did,

but people will never forget how you made them feel."

- **Accept.** Accepting feedback, different ideas, and other perspectives creates a collaborative environment in which everyone can thrive, everyone is given the latitude to explore, experiment, and make mistakes. The refinement process occurs the moment when leaders acknowledge their own vulnerabilities and blind spots.
- **Develop**. Coaching and helping others adopt a growth mindset is the key to an imperative described by James O. Rodgers in his book *Managing Differently: getting 100% from 100% of your people 100% of the time.* It is the determination and drive to develop other leaders. Great leaders always make you feel as if you are the most important person in the room. They are attentive, respond to emails, leave comments on your LinkedIn posts, and seize every opportunity to interact and build up their team.

L	E	A	D
Listen *Attentively*	Elevate *Others*	Accept* *Feedback*	Develop *Leaders*
Leverage *Storytelling*	Embrace *Values*	Acknowledge *Vulnerabilities*	Drive *Change*
Learn *What motivates*	Empathize *Show You care*	Address *Blind Spots*	Determined *Lead Differently*

1.1 Lead Framework **3 A's (Ask, Accept, Act)*

Great leaders proactively seek ways to make a difference in the lives of others by encouraging all around them to understand the *why* and not only the *how*. The why provides everyone with a sense of purpose and clarity, motivating people to keep going despite challenges or drawbacks.

The greatest leader of all time was humble, approachable, had great pity for others, and used questions to engage others. He even paid the ultimate price, giving his life to save the lives of many.

***

Authenticity

Authenticity in the workplace can seem only a nice idea because to survive in corporate America, one must adapt to the unspoken rules which govern an organization. It is difficult to be oneself when you are judged by your hairstyle, the way you speak, or your choice of entertainment. Authenticity is the ability to live by one's values while acknowledging the values of others.

Bringing my authentic self to work means I use my creative and innovative energy to lead with purpose and fulfill my personal purpose statement: "To use my H.E.A.R.T. (Honesty, Empathy, Attitude, Resilience, Time) to enrich the lives of others and help them reach their full potential personally, professionally, and spiritually." I welcome the privilege to coach and mentor people newly entering the workforce. I want to be a part of their leadership journey and be remembered as someone who made a difference in their lives.

Leading with Purpose

As one advances in their career, leading with purpose becomes a more critical part of success. Those who master the craft have people flocking to work for their organization and giving 100 percent. Winning the war on talent is not about how intelligent you are but how you lead and inspire; people like to perform and produce results for leaders who make them feel good about themselves and their work.

Leading with purpose is a choice, to succeed where others have failed, the choice to make a difference and inspire the people around us. This choice means we are more likely to be remembered even after we are gone.

Ludell F. McRae

Ludell McRae is a Senior Vice President, Program Manager at Truist Insurance Holdings where she oversees risk and compliance issues for the Insurance Technology Business Unit.

Learning from Leaders in Action

by Leonard Robinson

Prior to starting the ITSMF Executive Academy (EA) course, my favorite book on the topic of leadership was *The 21 Irrefutable Laws of Leadership* in which John C. Maxwell surmises that "Leadership is influence, nothing more, nothing less." Further, Maxwell points out that true leadership cannot be appointed or mandated but must be earned.

Because this was my go-to definition, when Dr. James Rodgers concluded an EA session with, "Leadership is selling an idea or vision in a way that people willingly follow," I was able to quickly agree; this seemed consistent with the idea that leadership equals influence. However, he also emphasized that influence is driven in larger part by the establishment of personal relationships and the exposing of personal vulnerabilities—not by demonstrating domain expertise, finding answers to wicked problems, or creating and communicating superb strategies. This was eye opening. This insight alone showed me why many brilliant minds do not realize their full potential climbing the executive ladder.

Here, I'd like to—based on this realization—focus on five skills/characteristics that seem to me to be critical for executive leaders.

<u>Authentic Character</u>

Authenticity of character was far off my radar coming into the course, but it is the trait which most resonated with me as we heard directly from leadership experts such as Marshall Goldsmith and Ed Welburn. During discussions, their dispositions seemed so genuine, without pretense or secondary motive, immediately inviting a level of trust. I left those sessions believing these leaders were sharing their unfiltered insights (truths) on topics in which they had both expertise and personal experience, for the sole benefit of helping me grow as a professional. When I compare the honesty and candidness of Marshall and Ed with the feeling of "being sold to" that I have often experienced in other such settings, it's easy to understand the large role of authenticity in being a successful leader. Authenticity engenders trust and the establishment of trust is critical to leading and influencing others.

Relationship Building

Relationship building for executive success is not just about how leaders foster relationships up their chain of command, but also peer relationships across the organization, and their subordinate relationships with the members of their teams. Relationships up the chain are critical for career growth; during transitions new leaders often bring trusted individuals along to new assignments. A promotion for the leader can mean a promotion for everyone in their close circle.

Building trusted relationships across and down the organizational chart are also critical. Several years back, Stephen Covey wrote *Speed of Trust,* arguing that companies could generate a competitive advantage by creating high-trust cultures which operated faster and more efficiently than competitors. Those kinds of cultures are built on trust, so leaders who establish strong relationships will be able to more successfully (and efficiently) execute strategies; in this way, building strong relationships serves as a competitive advantage.

Identifying and Fostering Talent

As a University of Alabama graduate, I am a big college football fan and often read articles about head football coach Nick Saban. In one, Saban told the story of when he first agreed to become the coach. On the plane ride from Miami to Tuscaloosa, Saban told his new boss, athletic director Mal

Moore, that despite being a great coach, Saban could not win games without talented players. Moore's response: "Thank God you understand that."

I share this story to highlight that great leaders cannot achieve great results on their own; they require great teams. The ability to identify and develop strong talent is critical. I take great pride in my own ability to spot and help advance strong talent. When I judge talent, I look at numerous factors including the ability to deliver results, solve tough problems, drive transformations, balance competing priorities, inspire others, coach and develop younger talent, and deliver in high pressure situations.

Because of my passion in this area, I tend to look at teams and leaders and assess how well they've fielded and developed others. In my 22-year tenure at P&G, I've worked on several outstanding teams, but there was one team that stood out. In 2011, I was Director of Analytics for P&G's Global Home Care category, and David Taylor was President/CEO of that business. During my time there, I was amazed at the amount of leadership acumen on that team—by far the most talented collection of executives I had seen in my P&G career (many of those VPs and Senior Directors would go on to become Business Unit CEOs).

In 2013, there was a race to determine the next P&G CEO in which the company placed each of its four Business Unit Presidents in charge of one the four business sectors. David Taylor was among this group of candidates, and although he was not given the best assignment it terms of business sectors, I was convinced he would inevitably prevail due to his ability to identify, develop, and surround himself with gifted individuals. This was a significant competitive advantage over his peers because it enabled David to spend more time thinking and planning for the future since he could trust the business leaders on his team to navigate daily challenges and successfully deliver the quarterly/annual results. In 2015, David was named the company CEO after delivering breakthrough growth for its health care business and he went on to have a very successful seven-year tenure leading the company.

While David delivered several transformations in his time as CEO, his most impactful change was restructuring the company into 11 individual categories and naming a CEO to lead each of those business units. This was a large change at P&G because we previously only had one individual with the title of CEO while other leaders were simply Presidents. However, David was able to operationalize his approach to leadership by identifying the top leaders and fully empowering those Business Unit CEOs to lead their businesses. By doing so, David also freed up his time as P&G CEO to focus on the future and ensure that he had a plan for the company to be successful for years to come.

Strong Communication Skills

Working at a large, matrixed company such as P&G for 20+ years, had taught me that clear communication is foundational to individual success; the ability to concisely communicate your need is the only way to ensure that your message is heard in a global atmosphere consisting of thousands of people, numerous competing priorities, and many organizational seams. My focus was placing a very high value on clarity of message.

However, during the early part of the ITSMF experience, we focused a lot on storytelling as a form of communication. The goal was to leverage personal experiences to develop an emotional connection with our audience. Soon it became clear to me that storytelling was far more effective than other forms of communication for inspiring people. While clear, concise words let people know what you want of them, it does not get into the why. And *why* is what motivates people to give extra effort. I now see that the best leaders effectively use stories that communicate the why behind the what to inspire action. As leaders we often need extraordinary efforts from our teams to deliver extraordinary results, so inspiring people with stories that connect with them on a personal level can motivate them to go above and beyond to reach our shared goals.

Grounded and Forward-thinking: Both.

A forward-thinking mindset enables us to paint a compelling vision for the future, something associated with strong leaders. During EA, we discussed the role of the CEO and what this person does every day. One of the key takeaways was that CEOs spend a lot of time staying externally connected to market trends to plan and envision the future direction of the company. This is critical for CEOs, but also necessary for all executive leaders. Change is constant and the best leaders are able to leverage their forward-thinking ability to anticipate and navigate change better than their competition.

The future will arrive, and forward thinkers will be ready. Meanwhile though, leaders are responsible for the outcomes of their strategies in current time, too. A.G. Lafley, former P&G CEO, in his book *The Game Changer*, says that "Execution is the only strategy that consumers see." Successful leaders stay close to operations to ensure they have the right systems in place for teams to effectively execute strategies; they also position themselves as resources to remove barriers which may prevent their teams from achieving the aligned goals.

***

While there are additional traits successful leaders share in common, if you can master the five above, you will exert a competitive advantage as you advance your career toward executive leadership.

Leonard J. Robinson

Leonard Robinson is Senior Director, Data and Analytics at Procter & Gamble. He has over 23 years of technology experience in a variety of IT roles.

Failure = Clarity = Empowerment

by Nick Sims

"Many are the plans in a person's heart, but the Lord's purpose prevails."

\- Proverbs 19:21

My journey to have clarity of purpose and to understand my core self has been riddled with failure, pain, and perseverance. At times I have failed in school. I have unintentionally failed people that I love. I have failed in business and many entrepreneurial ventures. My experiences living within broken relationships while determining my place in the world have been painful. These failures generated immense pain, but I know *I* am not a failure.

The scars from painful experiences are still with me; I feel them in memory and feel some physically. Each failure and pain have taught me perseverance. The gift from these experiences is clarity on what did not work, which provides a lesson to pivot to another path. As I developed a sense of self-perseverance, I learned there is joy in the journey. When I began to enjoy my experiences, I realized failure and pain are a welcome addition to my lifelong learning.

I am grateful for the pain. It has taught me to listen, pay attention to the world, and take pride in the successes that follow when you don't quit. Many of my early life lessons were learned in the school of hard knocks in the inner City of Detroit, Michigan. I am the son of an intelligent father who made one too many bad life decisions. I am the son of a crazy-smart and intensely strong-willed mother who is an entrepreneur. I am the product of the influence of the people I now dedicate my life to. And with this as my foundation, I am on the path to my unique version of success, moving with purpose on my journey through life.

***

Three pillars describe who I am: creativity, technology innovation, and empowerment. This core trilogy has become my sense of purpose, along with providing unconditional love and being of service to others is my mission (my *why*).

My vision (my *how*) is to be the best husband, father, son, and friend to my loved ones while serving the world. Creative innovation, thought leadership, and investment in the growth of people is my lens through which I serve. My values include creativity, love, passion, empowerment, and growth. My leadership principles are resilience, vision, authenticity, courage, and vulnerability.

It took me over 30 years to develop this sense of clarity and purpose without having a roadmap or a life guide. Additionally, I didn't have a strategy to establish a life plan. Through a combination of various experiences, a few books, and a few mentors who guided me through the process of thinking through my life, I've landed here.

With this as backdrop, I'd like to talk about my core pillars that serve as a foundation for my leadership.

<u>Creativity: my Foundation</u>

I've developed a quirky character trait: it is hard for me to function or think creatively without a white board or some form of blank canvas to shape thoughts into something tangible. For many years now, I've used the blank canvas to dream up the art of the possible. I am a true artist in the depths of my heart with passion for drawing, painting, music, sculpture, theatre, dance, and more. This passion for the creative arts stems from my older sister who engulfed me in music, her own passion for theatre as an artist/director, and a love of mathematics (an artform in itself!). Over my career, rather than creating pieces for public consumption, I've shifted the creative focus to drive other passions—technology innovation and empowering others.

During my time at Whirlpool Corporation, I led a team focused on technology solutions for the Global Product

Quality organization representing over 30 brands and products. Our function was keenly focused on increasing the quality of products to a best-in-class measure while also reducing the Total Cost of Poor Quality (TCpQ). There were several aspects, from a business, product design, and manufacturing perspective, to pursue this goal but it was my responsibility to determine how technology solutions could collaboratively support.

To develop the complementary technology vision, I spent time with several front-line operations teams and mid-level managers to understand the current state, learning about them and their passion for the customer. Equally I shared my own passion for the customer experience and how technology has made improvements in other areas of the business. With the foundation of an authentic and open relationship, we collectively brainstormed through various creative mediums: physical white boards; Miro, a digital white board; and rapid prototyping exercises. Various inputs from prior engagements, industry leading perspectives, and 1:1s with key functional executives provided the right insights that ultimately led to drafting a vision for the technology solution. At the epicenter was a visual representation of our customer with a roadmap that aligned with the functional strategy; the technology strategy became a component of overall business strategy. With that alignment, we were set to execute.

With the backdrop of my own digital canvas—which now comes in the form of an iPad Pro—and a whiteboard, plus a few sticky notes here and there, I can be the natural creative that I am.

Authenticity, Vision, and My Path to Value

I am authentic by being true to myself, my values, and convictions, and maintaining an authentic and transparent relationship with others. As a visionary, I seek to develop a clear and inspirational vision of the future, and the ability to communicate that vision to others who are willing to follow.

My core pillars and leadership traits tend to sync so closely together that it is challenging for me to completely separate them, along with the themes of entrepreneurship and empowerment. By understanding this about myself, I can frame my passion for tech with my start as an entrepreneur.

As a young Detroiter living in a single parent household, my first youthful stint as an entrepreneur came through manual labor by shoveling snow and cutting neighbors' lawns. It was evident to me that if you provided a high-quality example of what your services could provide, people were more likely to give you a shot. I did the best I could to make my own lawn and sidewalks look pristine to show the quality of my services. This act of self-promotion and a bit of storytelling opened the door for more business from neighbors and bled into other types of services. Before long, I was running the local paper route, washing cars, selling paintings, and generally stuffing my pockets with money by any means necessary. This led to two epiphanies: while I enjoyed the money, I wasn't fond of the manual labor because it was difficult to scale; and I so enjoyed seeing people happy as a result of my services that I began to seek that as an end in itself. Later, I saw this play out with my passion for technology innovation, which in my opinion relies less on physical labor than providing the outcomes of a better life, business, or process for others.

My passion has been refined over the years to one simple goal: tech that delivers value. My view of technology broadly spans science, engineering, software development, physical hardware, data and analytics, marketing technology, information architecture, user experience, physics, astronomy, biology, and much more. Although the world puts parameters on what is considered technology, the emerging advancements in the field are how I view the space.

To truly convert technology possibilities into something of value for consumption/use one must have the courage to risk it all, with a high propensity of failure. As previously mentioned, I am proficient at failing and failing toward the next failure! The pain and anguish from these failures requires resilience to keep moving forward to ultimately reach solutions that work. If you are a creator or entrepreneur, then you know this feeling all too well. Most people may not truly appreciate this feeling, and perhaps it classifies me as being a bit crazy, but it is also what drives the world toward new heights.

Innovation is at the heart of why I love technology. Innovation requires heart, a steady will of faith, and when done correctly, will result in happiness for those that find value.

Courage and Resilience

Courage involves making difficult decisions in spite of opposition and uncertainty, even when the odds are against you, while resilience demands the ability to bounce back from setbacks, overcome obstacles and persevere in the face of adversity. These two traits are linked and fuel my constant pursuit – empowerment of others.

In the future, when I'm 87 years young, on many days I hope to find myself sitting with one of the leaders of my YMCA speaking to a group of students about their passions. Some have come from battered beginnings and others simply need a place to unwind after school. Regardless of the scenario, my mission will be to be of service to these young people. By then, I will have generated enough wealth for multiple generations, positively impacting the lives of many others who are now doing the same. At that point, I am retired, yet in my constant pursuit of self-empowerment, still finding ways to let that passion bleed into the lives of others. If I can do this, I will feel that I've completed my life journey.

To realize that vision of service to others I seek empowerment now in every interaction. In November 2021, I was recruited to join Abbott Laboratories as the Global IT Director/ Business Unit CIO for Neuromodulation. I'll likely spend the rest of my corporate career at this healthcare company because the mission, vision, and values are aligned with my own. (Additionally, the convergence of technology and healthcare from the innovations of Abbott saved the life of a family member.) I know that the work that I do here not only impacts me, but many lives over.

Empowerment is keenly focused on the lives of people by providing them with the strength or means to be their best selves. I see my daily interactions with members of my team or with students as opportunities to tell my own personal story while inviting them to tell me theirs. Being vulnerable earlier in a relationship opens doors that can lead to a sense of security to have deeper discussions. It is this safe space that provides the window to empower someone else's life. The walls that shield the heart, erected from previous hurt, can come down when you are truly seeking to help. After taking care of yourself, be sure to put double the amount of time into other people. Be their bridge to greatness.

Nick B. Sims

Nick Sims currently serves as the Business Unit CIO, Neuromodulation, within the Medical Devices division at Abbott. He has more than 20 years of value-focused tech experience spanning industries including retail, financial services, high-tech, and manufacturing.

Leadership Takes a Village

by Floyd Springer

"A man with no country will always be lost! Never forget where you're from and who you are."
- advice from my dad after a few drinks following Sunday dinner

I have been blessed to live the dream my parents envisioned when leaving Barbados in the late 1970s, and moving to the US in the early 1980s, by way of Toronto, Canada, where I was born. They set the bar high when it came to taking risks as well as trusting their gut. At this point in my life and career, I can plainly see that many of their decisions to be bold have helped shape my own approach to leadership.

My family was living in Barbados when my uncle met a woman at a Jehovah's Witness conference in Florida, where he was living while contracted to cut sugarcane and perform other manual labor. On the third visit they married and moved to her hometown of Fort Wayne, Indiana. From there, Uncle Kenneth sponsored many of his siblings, of a family of thirteen, to relocate from Barbados, Toronto, and London to the state of Indiana.

Kenneth took on the responsibility of reshaping future generations of our family so that we would have all the opportunities that the United States offers. Military, college, and trade were all options my parents held out for each of their children.

Many of our family members had different religious beliefs, spoke with different accents, some Caribbean, some American, and others British. Yet no one held that against each other, nor made mention of it or thought anyone who remained "back home" in Barbados, or who spoke in a different accent, or had a different religion, was less than the other. That family dynamic modeled a true definition of inclusion and that is how I continue to lead. Like my uncle Kenneth, I want to bring everyone together from different backgrounds to improve their livelihood, families, and individual selves—including my wife, children, extended family, as well as friends, peers, and everyone in my organization.

Corporate Leadership Journey

"If you think education is expensive, try ignorance!"
- my dad (paraphrasing Robert Orben), who stressed education despite not going to college himself, when I complained about the cost of college

I started at Eli Lilly in 2001, after finishing a bachelor's in management information systems and a master's degree in telecommunications from Ball State University. I sought that advanced degree to ensure I had a great start in corporate America. I wanted to show my family that their hard work had paid off. I came into the job with all the energy and focus I could give, and arriving with a master's allowed me to start one level above others.

I was met with many challenges and stayed at that same pay grade for almost five years, while others who started with me, without an advanced degree, were promoted to my level and then passed me. After a change in leadership, I started to receive the gift of feedback and coaching. I learned that I was not asking for challenging work but was taking whatever others did not accept. I was seeking exemplary performance ratings but hadn't been assigned to exemplary projects. This all changed when I was given opportunities to work directly with manufacturing and quality IT organization.

Many other dads who worked in manufacturing in Fort Wayne took pride in it and spoke about their pensions with delight. I was excited to be back to something close to my hometown roots. But the main benefit of working with manufacturing? Feedback. Feedback was something I was missing early in my career and while I still worked in a corporate organization, manufacturing provided feedback early, often, and directly. It was a gift and a blessing, something that had been missing in my development. Without a mentor who knew my work, I had not gotten the necessary coaching vital to success. Working in manufacturing also allowed me to communicate with a business area, sharing benefits for each deliverable that required approved downtime to implement.

By 2006, we were building new facilities and growing. I began supervising contractors, then eventually Lilly employees, and after my first promotion three months later, I was approached about a formal leadership role. While I had some hesitation, my wife and I had three daughters ages two, three, and five, so this was not the time to turn down any opportunities to advance my career; plus, my wife assured me that I had all the capability to be successful.

Since taking that role, because of my performance, I worked with and met my first sponsor. Mentors can coach you, which is what the new leader in Telecom and many in manufacturing were to me; but a sponsor could open doors I didn't even know about. After working with her, I had several leadership opportunities at Lilly, and later she even brought me back to Lilly after being away for over three years. Living overseas in China for six months, obtaining my Six Sigma Black Belt Certification, managing global teams with different accents (like my family), traveling to meet face-to-face with staff and leaders in countries like Italy, India, France, and Puerto Rico—these were all amazing opportunities created by my sponsor.

Throughout that journey, I managed to be at every play, soccer game, track meet. My wife and three girls (now in their first, second, and fourth years of college) always knew I would show up. Leadership can be a challenge, but well-being, health, and community are equally important. I learned you must disconnect, recharge, and maintain balance; this allowed me to be at my best.

After a company-offered retirement program, I happily accepted an open director opportunity—where I could lead leaders. This was a role many of my coaches, mentors, and sponsors in corporate and manufacturing IT were at when I met them and, with their guidance and support, I had already listed the role on my career map. I learned quickly that the game was different at this level.

While I knew how to manage people and projects, managing relationships and supervision of my peers took a different skill set that I had yet to learn. The book *Quiet: The Power of Introverts in a World That Can't Stop Talking* by Susan Cain mentions that introverts collaborate with people they like and respect, while extroverts compete against those that they like and respect. I am an extrovert but as a manager with direct reports, collaboration was a plus and needed during times of chaos. Being part of a manufacturing site lead team, collaboration is a weekly, daily, at times even hourly activity—and early on I struggled. This was not due to lack of grit, working hard, driving—quite the opposite—but being vulnerable and asking for help from peers and leadership was a new skill I needed to develop.

<u>No one succeeds on their own.</u>
"And from this place of humanness, that can hold both strengths and weaknesses, we can do the most leaderly thing there is: Ask for help." - Peter Bregman

"Vulnerability is not winning or losing; it's having the courage to show up and be seen when we have no control over the outcome. Vulnerability is not weakness; it's our greatest measure of courage." – Brene Brown

I once took for granted that others would have a good feeling about working with me, took for granted the smiles on their faces, assumed an excitement to partner with me; I thought these were all natural aspects of executive leadership, or leadership in general. I've always been impressed by leaders who get things done with their own staff and exhibit grit. To my mind, being able to play quarterback, passing, running, and leading the team by "putting the team on your back," set leaders apart. However, what became apparent was that after I tried performing such feats, I was worn down and had no time for my own well-being. My peers and supervisors also did not have a clear understanding or buy-in of goals that they were not a part of developing or delivering.

The two quotes above were not within my view of executive presence or seen as leadership traits when first coming into my director role. Being open to request, accept, and recognize help is a process that requires both vulnerability and the willingness to partner with others to achieve goals. I've learned that leveraging your peers, supervisors, and sponsors, is as important and impactful as influencing your own organization to achieve goals.

Asking for help has been one of the most fruitful changes I've made recently. Not passively doing so but stating to my supervisor and/or leadership team: "I need your help to make this happen and be a success." I don't say this in every meeting, however when I do, I notice the level of engagement and care for whatever the deliverable is (or decision, strategy, organizational change), increases tremendously. And so does the probability of success.

Even though leaders want to be a part of something bigger than themselves, I've learned it is not required that we own it all. Being vulnerable, asking for help, acknowledging you don't know it all or have all the staff capacity and/or capability, shows humanness and humility. These are traits anyone would want to see in a leader they work for or with in any organization. They increase trust as this leader will admit when they don't have all the answers and create a collaborative environment *for all*, regardless of race, gender, religion, sexual orientation, age, making every team member feel included, engaged, and ready to succeed.

Advice for Future Leaders

"Always be yourself. Changing for someone else takes too much work!"
– my mom's soft but stern voice after I had a challenging day at school

One of the most important pieces of advice I can give is to seek out mentorship and sponsorship from individuals who can provide guidance and support. Being promoted is not a finish line, but a starting point; everything before was a warm-up in preparation. The next phase may involve reaching out to senior leaders within the organization or seeking out mentors and sponsors outside of the company. It is also imperative for Black executives to build strong networks of peers and colleagues who can provide support and advice; but that network should also be diverse versus everyone being similar.

Another key strategy for new leaders is to develop their leadership skills through training and development programs. Many organizations offer leadership development programs that can help build skills and competencies needed to succeed in more senior leadership roles. These programs may focus on areas such as strategic thinking, communication, and team building, and valuable opportunities for new leaders to enhance their abilities. (It was a sponsor who recommended me for several programs, including ITSMF.) Often, we are placed with other leaders during these programs, facilitated by experienced leaders. Learning the content is a small part of why you are in the class; the critical piece is networking and building relationships. Create cross-functional partnerships, provide help and assistance, seek out feedback, leverage your network of trusted peers as sounding boards and peer coaches.

New leaders, and indeed employees overall, need to advocate for themselves and their careers. This may involve speaking up about accomplishments and contributions as well as requesting impactful projects versus assuming your leadership will give them to you.

Lastly, be vulnerable, and open up to others. Personal relationships with colleagues pay back beyond measure.

Floyd F. Springer

Floyd Springer is Senior Director at Eli Lilly, with 20-plus years of experience partnering business and technology to deliver multimillion dollar solutions and services that add business value and meet industry-wide compliance expectations. His experience span includes companies such as Eli Lilly, Carrier, and Rolls-Royce.

II

Tools on the Ladder to Executive Leadership

Connecting with Confidence

by Lisa Brown

When I remember where my aspirations began, I think back to my childhood. As a little girl on the sunny Caribbean island of Trinidad I played on the beach, and with Barbies or GI Joes with my brother. During the week after school, I walked a few blocks in my stiffly starched Catholic school uniform to meet my mom, who worked in the accounting department at the phone utility company, located in what at the time seemed like a mammoth sized building in the small downtown area. There was a security guard in the lobby who would call her desk to let her know I arrived, and I thought Mom was the most important person in that building! That admiration must have seeped into my soul because I dreamed of working in a cool high-rise corporate building long before I knew what it would take to get there. I could not have known that years later in my own career, I would have the opportunity to sit at the table discussing my work with the CFO and CTO of a US Fortune 50 Bank.

I attribute my own journey to the executive suite to several factors. Work ethic, business acumen, and driving results were critical foundations, but I'd like to touch on three other areas that I continue to hone and refine—because in a field of talented hard-working people, these can be significant differentiators.

<u>Confidence. How do you show up?</u>

I am a Black female immigrant in America. Identifying in those ways comes with a bit of baggage—the history of this nation, the news cycle on TV, the corporate focus on diversity and inclusion all reminds us of the unique set of challenges that people like me face, and the potential obstacles, including the bias of others, which could potentially slow career success.

In my formative years however, I was a teenage girl in a country where being Black or being female didn't seem limiting: my Prime Minister, politicians, doctors, teachers, news reporters, mostly looked just like me. That shaped how I see the world because I was never told about limitations in my childhood. It never occurred to me that there was anything I couldn't be, so that's how I navigated the world.

Well known leadership author Dr. James Rodgers often reminds us, "Someone is going to doubt your abilities; don't let that someone be you." I hope that this resonates with you. Imagine for a moment how differently you would navigate your life and career if you shed some of the fear and the negative messages you tell yourself. Imagine how differently others would perceive you if you shed some of the weight of self-doubt. Lack of confidence is an absolute inhibitor of your potential.

In case you haven't heard it recently: you are smart, you are capable, you deserve to achieve your success destination, and there isn't anything you can't learn. That may sound patronizing, but over the years I've repeated many versions of it—sometimes to myself when I leaped into a new role that stretched me, and sometimes to young mentees or not-so-young peers who, despite their fancy senior job title, still needed a reminder that their success was well deserved and that their ambitions were achievable.

In a quiet moment (like right now), ask yourself, what is the negative message you've been repeating that stands in the way of your confidence? Things like: *I'm too overweight to be taken seriously. I don't have the same degree others have. I'm a single parent and others don't seem to understand my unique challenges. They don't invite me to happy hour because it's a boys' club. People don't ask my opinion because of my accent. I was passed over for promotion because I have a same-sex partner. I can't compete with younger peers because I need time to take care of my aging parents. They only considered me for this job to meet a diversity target. They have not considered me for the job because I am an introvert. They all remember my last mistake and I'll never again be seen as capable.*

Any of us could fill pages with self-doubt statements. And yet, if asked to provide the hard evidence that makes the statement true, most times we cannot, and thus become the person doubting our own ability. Do whatever it takes to move toward confidence. Some ideas are: morning affirmations, talks with your pastor, counseling, healthy eating, and exercise, even the fake-it-till-you-have it approach to confidence. However you tackle building confidence, it is critical to unravel the negative self-talk that could hold you back from making important life and career decisions.

Confidence needs to be the armor that you wear to work each day. Dress confidently, walk confidently, make eye contact confidently, greet people confidently. When you walk into that conference room and take a seat at the table, don't slump your shoulders and mumble your words. If you are disinterested and disengaged, and internally questioning why you are here and whether you are worthy, everyone else may do the same. If you walk in with your shoulders held high, wearing a winning smile, and thoughtfully ask your questions and provide your insights, others' perception of you will be different. I guarantee it.

Confidence is not always innate. But the best news is that I've also improved my confidence through observation and practice. It can be learned!

<u>Relationships. Who knows you?</u>

Now that you have a little boost in your confidence, it's time to put it to work. I'll avoid the dreaded overused *networking* word, and instead talk about relationships built on shared interests, trust, respect, and understanding. If you are truly excited about your company/industry, you will have thoughts and questions swimming in your mind, and although your neighborhood friends may not be interested in hearing you drone on about interest rate spreads, agile transformation, cloud migration options, and artificial intelligence ethics, other like-minded leaders will.

I encourage you to take the leap and schedule a conversation with that person whose work or leadership style intrigues you. You may be surprised by people's willingness to share their work and their journey, and to learn about you, your goals, and your ideas about the business. Equally important in finding leaders that you admire and respect is that as these relationships evolve. If you find one or more of these folks would be suitable and willing to be a mentor who provides guidance and feedback, or a sponsor who is willing to advocate for you behind the scenes, and to open their own networks to you, be sure to nurture and develop that relationship.

By building and sustaining meaningful relationships, I've navigated from sales, to strategy, to operations to technology, often without a formal application. Yes, I learn new things quickly, but without those coffees and lunch meetings and hallway chats, I would not have learned about jobs before they were posted, budget cuts and reorganizations before they were formally announced, meetings and happy hour events that weren't on my calendar, the inside scoop on talent conversations when I wasn't in the room, and cultural faux pas I should avoid. All of that information helped make me a more informed leader, who made better decisions, which brought value both to the firm and to my career.

If you have not done much by way of intentional relationships, start by making a short list of people you admire and would like to get to know. I generally try to maintain relationships with subject matter experts in emerging areas of interest, business partners who can give me a broader view of the enterprise (if you are in technology, those partners may be in finance, operations, sales, human resources, or risk, for example), my manager's peers, former bosses/co-workers. For extra credit I try to schedule a coffee chat with anyone that I've had a difficult working relationship with recently. Often, I leave the discussion with a clearer understanding of the other person and what led to the difficulty, and sometimes we agree to disagree, but either way that chai tea latte connection means I've learned something about that person and how they think, how I am perceived. And we've both gotten some caffeine.

If you already enjoy building relationships, then review your list and ask yourself these questions. *Do I have meaningful relationships with diverse people who bring varied perspectives? Can I confidently name at least two mentors and one sponsor? Do I have meaningful relationships with people in my industry outside of my current company? Are my relationships balanced with people who are junior to me/at my peer-level/more senior? Do I know what is being said about me at the leadership table when I am not in the room? Am I inviting and being invited to non-work functions that demonstrate a deeper than surface level relationship? Have I had a touch-base with most of these people in the past 12 months?* If the answer to any of those questions is no, then you too may have an extra credit assignment worth considering.

There are rarely true self-made leaders. Everyone I know can attribute their success to a mix of preparation and hard work, a few lucky breaks along the way, and someone who gave them some critical coaching, listened as they practiced a presentation, or offered a chance to do something new. Take the action to honestly evaluate your current relationships and decide how to best invest further in nurturing them.

Vulnerability. Do you ask for help?

While you are refreshing your confidence and building stronger relationships, another critical trait/skill is needed—a healthy dose of humility and the ability to be vulnerable.

I joined the banking world through a leadership development program right after completing an MBA at Emory University. Less than a year later, during a rotational assignment geared toward client satisfaction, someone in the group left. The regional executive said something along the lines of "You seem pretty smart, I think you could help fill in for a while" and so began my new role as market manager, overseeing 14 financial centers. When I showed up, bright-eyed and enthusiastic, I could see the evident worry on the teams' faces. *What does this young woman, who has never worked in a bank branch, know about running this business? And why wasn't John Doe, with years of experience, given the opportunity?*

As I sat with each financial center manager, I think they were surprised by my approach, which was to reassure them that they were the expert in the day-to-day business which I needed help to understand, and that my job was to listen, learn, help remove obstacles to success, identify trends in reporting that could pinpoint areas of opportunity, and that together our goal of driving successful sales, service, and operations could be achieved.

Disarmed by that approach, a few were comfortable to share that they were not strong at reporting and data analysis, and that I could help them in that area. A couple of teams said they really were struggling with sales and hoped that my prior client experience work could help improve customer satisfaction results.

While confidence was crucial in accepting the role and believing that I could perform it successfully, vulnerability was equally important in building trust with the team. The teams saw that I was honest about my own blind spots, and I think that in some ways made them more open to the idea of driving success together.

Often in a new role we feel compelled to immediately show how much we know in order to establish credibility, and we continually re-affirm that position by never letting our guard down enough to ask for help. I have found that setting down the superwoman cape can create a different team culture where teammates feel empowered and inspired to add value and to be transparent about their own vulnerabilities. Not asking for help often creates a culture where no one else asks for help either; as a result, you may all end up drowning together! Listening, learning, and finding the right moments to show that you are human all go a long way.

Another way to demonstrate vulnerability is by sharing your own journey. For a very long time, I felt that my immigration story, my parents' divorce, my full-time job to pay for college, my feelings about racial tensions in the country, or my ongoing struggle to balance family/work/health priorities, all should not be discussed at work. I believed that these topics were things that other leaders didn't worry about, that they might make me appear weak, and most importantly, had nothing to do with the critical fiscal planning or technology release at hand.

But as it turns out, there are immigrants everywhere. There are divorcees and single parents. There are people who are currently working to complete a degree. There are people who are having a tough time focusing on work because they cannot fathom how, after a widely publicized shooting the day before, people could come to work and continue business as usual. There are women who are skating into work late, sweaty and exhausted at 9:05 because of all it took to get their children ready and onto the late school bus that morning. And it turns out that when I let my guard down and share my own stories of life and imperfection, all these other people then feel safe to share their own experiences and challenges on and off the job—and that brings a team closer together and drives even greater productivity.

I don't believe that there is only one path to executive leadership. I have witnessed many recipe combinations that led to success. The three pillars I described are a few ingredients that I see consistently in leaders I admire, and that I have tested successfully in my own career journey. Your path will be unique, but with inner-confidence, meaningful relationships, and a healthy dose of vulnerability, you are well on your way toward being a better leader and growing toward your definition of success.

Lisa J. Brown

Lisa Brown is a technology executive who leads the Enterprise Data Management Services Organization for Bank of America. In this role, her organization is responsible for the design, development, and maintenance of Enterprise applications.

Know Yourself, Know Your Leadership Philosophy

by Jonica K. Cowan

It was in December of 2019, at my organization's training for new senior managers in Las Vegas, Nevada, where I received my first lesson in leadership. After we made a red-carpet premiere and received congratulatory well wishes from our leaders, we listened intently as the host explained how our job performance as individual contributors allowed us to ascend to the role of senior manager. He also reminded us that our journey was just beginning and that during this next phase, we would no longer be judged by our ability to perform on an individual basis. We would now also be evaluated on our ability to get work done through others.

Throughout the remainder of the training, I began to think about my approach to leadership and the type of leadership skills I wanted my direct reports to see in me. Prior to that, I used to think that my ability to do and get things done as an individual contributor would be all that was required to the next level, that my contributions and work ethic would allow me to be seen as someone ready for a position in executive leadership. But as Marshall Goldsmith points out in his book, *What Got You Here Won't Get You There*, this idea required reconstruction. I needed to shift my thinking and actions from *doing the work* to *getting the work done through others*. I had to stop and focus on being the person that guides and inspires those under my charge to willingly follow me and encourage and empower them to excel at doing the work themselves.

First, I would need to decide how I want to lead, the principles or values I will use along the way, and the audience that I will focus these traits on going forward. Second, I would need to utilize these traits to craft and adopt a leadership philosophy statement, idea, or premise that defines my leadership strategy. The statement would be based on my personal experience and previous interactions with various leaders and would also need to clarify what followers should expect from the philosophy. The actions in this statement should be transparent, consistent, authentic, and intentional. Finally, as I began to craft a leadership philosophy, I knew I had to focus on using knowledge of self, servant leadership, and staff development as the foundation. I also understood that I should continue to rely on my own personal experiences, the principles that I have learned through self-discovery, and observations of those principles in others I've interacted with along my journey. (I'd also pay attention to experiences that showed me what traits exhibit toxic leadership, traits that do not inspire willing followers.)

I have learned that it is the understanding of oneself that will aid in my success as a leader. My ability to understand my character, behaviors, motives, and how these affect my ability to lead others forge my ability to make consistent and effective decisions. Being able to identify and manage my emotional triggers allows me to manage decisions and relationships. Continually assessing myself and looking for ways to improve and grow as a leader enables me to make better choices that will help me further develop—and also allow me to make choices in the best interest of my team. Mahatma Gandhi once observed, "The best way to find yourself is to lose yourself in the service of others." This advice helped me understand that placing the needs of others before my own is an important step.

Being a leader is more than managing people and ordering them around just because I am in charge. If I am going to be a successful leader, I must first be willing to serve, and put my needs second to the needs of my team. This approach lets my team know, undoubtedly, that I will always be there for them and have a vested interest in their growth even if it means that they move on. I must be willing to see situations from their perspective, make decisions with their best interest in mind, and make sure they have the resources they need to successfully achieve their goals and objectives.

I used to think that the sign of a great leader was equivalent to the size of their followership—Dr. Martin Luther King Jr., Mahatma Gandhi, and Nelson Mandela all had large followings. As they dug deeper to better understand the goals and objectives of their followers, they were creating more leaders to support their cause. They each led by example, continued to put others before themselves, owned their mistakes, and empowered their followers. My job is now to do the same for those who are under my charge. I must seek their potential, nurture it, and provide them with opportunities to achieve their goals. To do this, I need to lead with a clear vision, provide concise direction, and coach others to perform at higher levels.

In the ITSMF Executive Academy, I learned that leadership is not a "one size fits all" scenario nor is there a "secret sauce or award-winning recipe." Instead, we all need to take the necessary time to develop our own views and definitions of leadership.

Effective communication is also woven into my leadership philosophy—equipping me to lead my teams and others in the organization to success. Communicating effectively is not so simple, and can be difficult at times, more complicated than the basic tasks of sending and receiving information. My ability to harness good communication skills aids in translating important data quickly and accurately, and proves essential when interacting with the team, understanding their goals, and resolving grievances. Skillful communication is also key in fostering a rapport with my team that would support an increase in productivity and efficiency. My communications must be compelling and persuasive. When leading a team through change and uncertainty, I must communicate in ways that build a more resilient workplace. This enhances my ability to interact more efficiently, share the team's vision, and build meaningful relationships.

Any skill that is learned requires a strategy to maintain and eventually master that skill. Communication skills demand practicing active listening, to gain a clear understanding of another person's point of view. Active listening fosters mutual understanding between speaker and listener, limits misunderstanding, and introduces clarity of thoughts, and variance in perspectives. To master this skill, I'm required to listen with intent, to capture the content and feelings of the speaker, thus enhancing trust, establishing empathy, creating transparency, and fostering loyalty. Intentional effective communication tools foster the environment for open door approachability and honest dialogue.

A personal leadership philosophy is just the starting point, a living document that serves as my point of reference in making both big and small decisions on behalf of my team. I must continue to evolve this philosophy as I gain new experiences and insights, using it as roadmap for how I will lead when interacting with others. Remaining true to these values and principles, I can work towards making an impact on those under my charge and my peers. Because in the end, leadership is not defined as a position or a title, but by the people I am leading.

Jonica K. Cowan

Jonica Cowan is Deputy Business Information Security Officer at Deloitte. In this capacity, she drives security strategy development and implementation globally for eight business lines supporting 350,000 global employees.

New Executive Roles Require New Mindsets

by Cynthia Dixon

Parts of the following include fictionalized examples of people in leadership roles that capture lessons from the ITSMF Executive Academy experience. Similarity to real or fictional characters or organizations are unintended and coincidental.

In corporate-speak, a "doer" tackles "hands-on" work and direct involvement in a product or service, while a manager is responsible for the day-to-day operations and ensuring the work is done. An executive is typically accountable for an organization's overall success and focuses on the big picture—strategic decision-making. Professionals in all three roles can be leaders; however, the higher a person advances in an organization, the less he or she "does," and the more strategic, influential, and accountable they become. The role of an executive leader requires a new mindset.

I started my career in technology as a doer, programming and coding software to deliver functionality. I remember a sense of great professional fulfillment when I wrote my first program and saw my first product into production. The code reviewer complimented the logical way I structured my code, the users were excited when the code delivered the functionality they needed, and I was beaming with an intense sense of accomplishment and pride for creating something tangible that brought joy and business value. As my career progressed, I continued learning and gaining expertise, doing and delivering more.

After several years, I became a manager and no longer coded or developed programs. "Doing" was over. I managed the day-to-day operations of the team to support the business and empowered my team to work toward a common goal. Still, I remained close to the work to ensure it was done, and when necessary, rolled up my sleeves and jumped in to assist, getting that same sense of accomplishment.

Next, I transitioned to an executive role and of course my accountabilities changed to being more focused on the big picture, setting the direction and vision for the team, and inspiring them to work toward a common goal and deliver results. This proved to be another rewarding experience.

In his book *What Got You Here Won't Get You There,* Marshall Goldsmith shares insights about how, often, the workplace habits that bring a leader success early in their career may actually delay career progression if the executive does not adjust. Thus, the skills that brought me success as a doer would not make me successful as a manager. And the skills that made me successful as a manager would not make me successful as an executive. While I continuously built upon the experience and skills from each role, executive leadership required a different mindset, a higher level of thinking about influence and accountabilities.

Understanding the business

An executive leader understands the organization's business strategy and how to support it—the internal connections and dependencies and the external factors too. Macroeconomic, geopolitical, and societal issues are constantly changing and often affect business, so leaders must read, staying abreast of current events, touring business operations, and connecting with other leaders and experts (internal and external). Being aware of what is going on around the world and how it impacts their business is smart leadership and essential to making effective business decisions.

Blockbuster, BlackBerry, and Kodak are examples of how companies that once led their respective industries failed by not understanding and adapting to changing market dynamics, their downfalls attributable to a combination of internal and external factors.

Blockbuster, the dominant video rental chain, failed to recognize the emerging shift in consumer behavior towards online streaming and digital content. While Netflix was disrupting the industry, Blockbuster clung to its outdated brick-and-mortar model, missing the opportunity to pivot and embrace the digital revolution.

Similarly, BlackBerry, the pioneer of smartphones, fell victim to its inability to keep pace with evolving consumer preferences. As competitors like Apple and Samsung introduced touchscreen devices and app ecosystems, BlackBerry failed to innovate to meet the changing demands of smartphone users, leading to a rapid decline in market share.

Kodak, once synonymous with photography, found itself on the wrong side of the digital revolution. Despite inventing the digital camera, the company was hesitant to fully embrace the technology, fearing it would cannibalize its lucrative film business. This shortsightedness resulted in Kodak losing dominance as digital photography became the new norm.

These failures serve as constant reminders of the need for leaders to remain vigilant, continuously innovate, and understand, and address the evolving needs of the external market.

Inspiring and Influencing Others

Although executive leaders are not responsible for day-to-day operations, they are held accountable for the organization's success, which hinges on the ability to influence and inspire others to achieve remarkable things. To maximize their teams' potential and inspire them to work toward a common goal, leaders provide a clear vision, guidance, and the necessary support to allow teams to flourish.

Dr. James Rodgers, a renowned executive coach with whom I had the privilege to work during the ITSMF

Executive Academy, shared two statements that continue to resonate with me.

- A leader must provide clarity of their vision. People follow leaders who know where they want to go.
- A leader must show vulnerability. People follow leaders who make it clear the leader needs them.

Though leaders may have different leadership styles, that style can have a significant impact on their team's performance and morale. A leader's style can generally be either transformational, transactional, authoritative, collaborative, or servant, yet the best leaders learn to flex their style. Here is an example of a situation that required a leader to flex.

Dan had an authoritative leadership style. He believed the best way to get things done was to be direct and assertive. At one point, his organization was tasked with a difficult project, and everyone was feeling overwhelmed and stressed. Dan's normal approach to pushing his team members harder was not working. Morale and productivity were quickly declining, spurring Dan to change his approach and he called a meeting with teams across his organization. Speaking with empathy and understanding, he acknowledged the challenges and asked for thoughts on the best way to move forward.

This change in leadership style inspired everyone to work together toward success with a common goal as Dan guided, supported, and empowered his team. His ability to flex his style overcame a challenging situation.

Although Dan had the title, he learned that having a title does not ensure people will follow. In the words of Brian Tracy "Become the kind of leader that people would follow voluntarily; even if you had no title or position."

Building Relationships

The power of relationships is so important to professional growth; no one achieves success alone. Cultivating relationships is fundamental to driving collaboration, innovation, and achieving shared business objectives.

Mark, a visionary executive leader, once observed that teams across his organization did not work well together—barely making their annual goals and yet they were challenged to increase results by 10%. Recognizing the power of relationships, Mark set out to build bridges across teams.

He began by organizing cross-functional workshops, bringing together employees from different departments to collaborate on projects. These interactions encouraged open communication, fostered empathy, and built a sense of camaraderie. Mark also initiated team-building activities and social events such as pickleball, volunteer opportunities, and networking events so employees could connect on a personal level, breaking down barriers and strengthening relationships.

As trust flourished, Mark harnessed the collective power of these relationships to drive innovation. He facilitated brainstorming sessions where ideas flowed freely, leveraging the diverse perspectives and expertise of team members. This collaborative approach resulted in groundbreaking solutions and enhanced problem-solving capabilities. Teams that were once isolated now worked seamlessly together, united by a common vision. Productivity surged, employee morale soared, and the organization exceeded its target.

By fostering connections, Mark unleashed the full potential of his workforce, creating a culture of collaboration and empowerment. Through his dedication to building relationships, he not only achieved success but also cultivated a harmonious and thriving organization.

Relationships also impact individual growth. You and your team may be doing a stellar job, but if the right people do not know you, your career and organizational impact will be limited. To grow, one must build relationships, promote a positive brand, and gain exposure within your organization.

Kay, a high-performing leader in a large organization, was smart, strategic, and always went above and beyond to deliver strong results. Yet despite her best efforts and impressive performance, Kay found herself constantly passed over for promotions and opportunities. She had all the skills and qualifications to perform in more senior roles and could not understand why her career was stagnant. One day Kay was discussing this situation with her mentor, and she realized the problem: no one knew who she was.

Kay had been so focused on her work that she neglected to build relationships with her colleagues and other leaders across the organization. She assumed her performance would speak for itself but now realized that it was important to network and connect. Kay made a conscious effort to turn things around, attending company and networking events, volunteering in company-sponsored community activities, and connecting with colleagues and leaders across the organization. Over time, Kay was promoted and awarded the opportunities she had been seeking. Kay learned that performance is important, but the right people must know you to advance to the next level.

Cynthia D. Dixon

Cynthia Dixon is Senior Director, Information Technology at PepsiCo. She is experienced in leading sales, go-to-market, finance, and digital transformation while passionate about leveraging the diverse skills and talents of people, combined with the power of technology to deliver the highest level of services for the business, consumers, and employees.

Five Slices of the P.I.E.

by Desmond Ebanks

What is executive leadership? Why would anyone want to become an executive leader? Is this something you have to do at a certain point in your career, or something that just happens (if you are lucky) after 20- or 30-plus years with a company?

I start with these questions because I must confess, prior to completing the ITSMF Executive Academy I had at best, a rudimentary understanding. So let me start by providing my own answers to my own questions.

Executive leadership is the ability to function at the top of a business and set a direction that others are willing to follow. One must possess highly desirable and transferable skills that enable performing in many different organizations. And none of that happens organically, or just because you do your job well. There are specific choices you must make and steps you must take to become an executive leader. That path is different than the path to becoming a valued subject matter expert (SME) or a great people manager. My own path to executive leadership moved through a specific set of principles—five main ones.

Principle 1: Communication

In elementary school, I won a writing award by making up a grand story about a picture of a monkey getting out of limo wearing sunglasses and jewelry. In middle school, I was captain of the debate team and played Daddy Warbucks in our school rendition of the musical *Annie*. I served as class president in sophomore year of high school. In college, I worked as a resident assistant (RA) and student orientation leader—making parents feel comfortable sending their students away for the first time.

The common thread? Communication. Written, verbal, physical. Even while a student, I connected success with the ability to communicate convincingly.

Through a series of fortunate accidents, I discovered the career of technology consulting. My first job was with PricewaterhouseCoopers (PwC) and two years later I joined Andersen Consulting (now Accenture). I would soon come to learn that communication is also key in business.

My original area of mastery was relational databases. I understood intuitively how to take objects in the physical world and model them in relational database form. In several early consulting roles, I created presentations and database diagrams explaining the back end of complex solutions. I picked up Structured Query Language (SQL) quickly and used it to explain the state of things we were building and systems we were testing, eventually becoming the Management Information Systems (MIS) guy—"the Answer Guy"—who could explain in detail why and how we were succeeding or failing at specific tests we were running on our system. The audience for my presentations continued to grow until I found myself in front of more than 100 developers, analysts, clients, and Accenture Partners.

This is what I call "Bottom Up" communication—summarizing details so they are consumable by a broad audience. I learned the technical language to find the data, and I learned enough business language to explain to those who only cared about the overall success of the project.

I learned "Top Down" communication from a Partner at Andersen Consulting in the late 1990s. Prior to this point I created many PowerPoint presentations and believed I was good at it. In preparation for an early morning client presentation, I spent several hours with one of the most senior partners at the firm in a Kinkos store adjusting fonts, aligning objects, and changing colors.

This hours-long effort was not at all about the detail. The detail was right days earlier. This was about ensuring that the presentation flowed without any distraction. The experience solidified the idea that if a multi-million dollar partner was spending time with me adjusting fonts and moving boxes, the effort was worth doing. "Top Down" communication is the ability to create a 10,000-foot picture, minus tiresome details, and still tell a compelling story.

Communication is the skill that lets you stand out from the crowd at the start of your career and will absolutely be essential as you reach for executive leadership.

Principle 2: Performance / Tooting your Horn

If professional growth and executive leadership in particular is a game (and I agree), then performance is the table stakes. None of the rest matters if you cannot demonstrate stellar performance; you must master delivering within the confines of your "Day Job."

Be clear about your goals for the year and ensure you and your manager agree on the definition of both meeting and exceeding expectations. Each goal should have an associated metric to measure results. Work smart to exceed these goals and then get comfortable with sharing your achievements.

A mentor once told me, "Toot your own horn, because no one else will toot it for you." This is good advice. Do not be shy about sharing your achievements, and always be generous with sharing credit. Very few accomplishments are delivered alone; give your team credit for participating and contributing to your success.

Principle 3: The Rest of P.I.E

In *Empowering Yourself: The Organizational Game Revealed*, Harvey Coleman claims the key to managing your career growth is awareness of three components: Performance, Image, Exposure (P.I.E.). What might be surprising is that he says performance is only a one-tenth slice of this P.I.E. Image is 30% and exposure is 60%.

Those communication skills will come in handy when working on your image and exposure. You can help manage your image by strategically sharing your accomplishments with your managers, peers, and in skip-level meetings, where others form an opinion of you. Participating in outside events and building connections with company leaders will also bolster your image and give you exposure to executives you might not otherwise meet.

Making personal connections is a learned behavior. Getting to know professional acquaintances and strangers better is a matter of discovering differences and similarities through conversation. This is a skill that gets better with practice. You may choose to pursue sport or hobby activities, golf for example, with the intention of having more things in common with the executives with whom you want to build relationships. You may be surprised to find you have more in common than you expect.

Deepening those relationships can ultimately build sponsorship. Coleman points out, "It is not who you know, but *who knows you.*" The greater your exposure, the more well known and appreciated you will become.

Principle 4: Mentors and Sponsors

A personal board of directors is a group of people who are close to you, who will give honest feedback, and whose opinions you trust. Forming this group is important; it may consist of family, friends, and former bosses. In this group is a subset you may call mentors—people who can advise you on personal and professional decisions, people you can trust because they have your best interests at heart in all areas of life. If they work in your company, they may be senior executives who have been in your shoes and are willing to share advice on how to navigate challenges. You can choose a mentor, simply by asking. You may not need to say, "Will you be my mentor?" but you can say something like, "May I put some time on your calendar again to discuss my progress?" With time, the relationship will likely take hold.

Sponsors are a different matter. You cannot choose a sponsor; a sponsor will choose you. How? If you are managing your P.I.E., while building professional relationships, a sponsor will notice, and emerge. For example, when meeting with a mentor, you may ask, "Is there anyone you recommend I speak to for advice on this topic?" And you may then ask that next person the same question. The result, over time, is you become known as a high performing person, with a broad circle of relationships. A sponsor will then be willing to spend his reputational capital on you. This is someone who is senior enough to place you in a desirable role because your performance and image will give them the confidence to do so. They find you through your attending to your own exposure.

Principle 5: Confidence

There is nothing more attractive than confidence. Confidence draws people to you and helps you build allies, and also helps you build high-performing teams. Confidence will help you manage your image and increase your exposure. So sure, confidence is great, but how do you get it? Practice.

All five of my principles are learned behavior; some may come more easily than others, but all can be improved with practice, especially confidence.

Well-honed communication skills lead to confidence. Practice by accepting invitations for public speaking, joining an organization like Toastmasters, and making professional presentations. Use your personal board of directors for feedback in these scenarios, gathering honest opinions. Take their recommendations, try again, and practice until you feel the confidence.

Confidence also grows via mastering your work subject area. Aim not only to complete work tasks but take the time to study and seek additional expertise on the subject matter. YouTube, podcasts, and audiobooks are great on-the-go means of enhancing your level of understanding of your job. The more you know about the work you do, the more confidently you can perform that work. That confidence will show.

Confidence is understanding that you have what it takes to become an executive leader. There is no special school, zip code, or gene required. Each of us can apply the principles I have outlined and move into executive leadership. Executive leadership is not a destination, but a journey. Perhaps reading these words is your first step.

Desmond Ebanks

Desmond Ebanks is the Technology Executive for Operations Technology Process Automation at Bank of America. He manages a global technology delivery capability dedicated to implementing Robotic Process Automation (RPA) and Intelligent Automation (IA) solutions throughout the organization.

Fueling A Career with Feedback – In all Forms

by Fred Love

When I was a kid, I loved building things, starting when I was five and got my first set of Legos. My imagination went wild with all the things I could create. I made buildings, cars, trucks, airplanes, spaceships. (I even made abstract art, and when I showed my mother, she'd say, "Hmm…that's interesting. I liked the airplane better.") After a few years, and a few more Lego sets, I moved on to building models. These were the old plastic model kits where you had to lay out all the parts and use special glue to connect everything together in the correct order. I built cars, airplanes, and spaceships with moving parts, then painted them and created displays using milk crates.

Then came video games and my eyes were opened to the worlds and game play that could be created with computers. In high school I was introduced to computer programming, and moved from basic programs to answering math problems, then creating cool graphics and entire video games. When my father got me a computer, I would spend hours writing programs at home. There was now another whole new world where I could create almost anything I could imagine. Sometimes I would not hear Mom telling me to go to bed so she would just come into my room and turn the lights off.

It was this passion that led me to Rensselaer Polytechnic Institute where I majored in Computer Science with a dual minor in Electrical Engineering and Business Management (I was blessed to have a five-year scholarship). While there I interned with Hewlett Packard. During those summers I improved my software development and electrical engineering skills. But I also noticed that the full-time employees I was working with had a certain level of business acumen that I lacked. Hence that second minor in business.

I eventually took a full-time role at Hewlett Packard and after eight years was offered the opportunity to become a leader. I was quite happy doing engineering work but my older sister, who was climbing the ranks in her career in advertising, encouraged me to take the offer, to advance my career. I had a great rapport with the team I would be leading and even had conversations with them about me being their manager, and they seemed encouraging.

As I progressed though my career journey there have been lots of lessons, some learned from formal leadership training, books and articles, and TED talks; some from coaches and mentors; some from personal experiences and overserving the experiences of others. The few stories that follow highlight leadership principles that have helped me to succeed.

Stop doing, start leading.

After advancing as a leader in Hewlett Packard and moving to Capital One where I became a Director, I felt as if something was slowing me down. A few of my managers hinted at what the issue was but I think I tuned out this feedback because I just did not want to accept it. And then came the hard realization that I had to "stop doing and start leading."

Wait. What? Going back to the age of five, I was the one *doing*. Building and creating with those Lego blocks. Building models. Creating computer programs. Software Development at Hewlett Packard. Now I had to face the harsh reality that for me to be successful as a leader, I had to focus on *not* doing. I had previously seen my value from the perspective of delivering a product, a service, an application, a process, or a major program. *Doing* the work had become a comfortable place where I had direct control over the outcome and took pride in being able to say, "I built that" or "I created that."

But it became clear to me that when you become a leader, you must transition from doing to leading. Just to be clear, by being a leader you *are* doing something, but it's no longer the expectation of your job responsibilities to directly produce anything. Your job is to effect outcomes by supporting, influencing, directing, empowering, and inspiring others. It was hard for me to hand off my very last program. I remember trying to give advice to the engineer who would take over that program. He looked me in the eye and said, "Fred, I got this." And he did have it. Though I had achieved a lot with that program, he took it to the next level, expanding the scope and making an even larger contribution to Capital One than when it was in my hands. As a result, I had the additional bandwidth to *lead* my organization to drive larger initiatives with even greater impact.

<u>Give feedback like my seven-year-old.</u>

Sometimes you learn some great leadership lessons from the youngest around you. One afternoon, I had gotten so focused on work I did not pay attention to the time and my wife reminded me that I had to pick up our daughters from school. I quickly grabbed a sweater, jumped in the car, and sped off. When I arrived, the girls, ages seven and ten, were playing with some friends so I took a moment to talk to one of the other parents. The girls noticed me, picked up their backpacks and approached me. My seven-year-old greeted me with, "You're late" using her outside voice. Then she pointed directly at my stomach and said loudly, "Your sweater has a hole in it." Now that's direct feedback.

As a leader I have found it extremely important to give similarly direct and unambiguous feedback. That's easy when it's positive feedback. But many leaders find it difficult when the feedback involves performance gaps or news the recipient might not want to hear. What's also key is being timely with feedback. Early in my career, I once had a manager who gave me feedback in my end-of-year review about something they had an issue with that I had been doing for six months. When they told me my immediate response was, "Why didn't you tell me six months ago when this first happened so that I could have made the correction?" At that moment I decided that once I become a leader, I would communicate feedback *when an issue occurs* so the person has ample opportunity to address the issue, and then, at review time, have a positive story to tell about how they turned things around.

When my daughter told me that there was a hole in my sweater, I was embarrassed but also glad; the hole was a blind spot. Likewise, I have so appreciated feedback received from others during my career journey that helped me to identify and address some blind spots. At the companies for which I have worked there is a strong focus on getting feedback during regular review cycles at mid-year and the end of the year. But I encourage my colleagues, team, and leaders to provide me feedback at any time, so that I can make necessary ongoing course corrections.

I have appreciated the value of having a strong network, people I trust and respect who provide direct feedback with candor. I do love hearing positive feedback but even when it's negative, I have learned to receive it with grace and just say "thank you." Less than positive feedback is good as we often have blind spots we don't see. If you hear from two or three people that you are doing a great job in certain areas, that's terrific. But hearing from two or three people that, for example, your communication style "rubs people the wrong way," may be hard, but you need to accept it and consider what you are going to do about it.

<u>Be relatable.</u>

A factor that enhances leadership is your relatability, which is directly tied to transparency. You can have the greatest vision or strategy but if people can't relate to you or feel as though you are not a transparent leader, you may not have their trust. It's key for people to know your story. Sharing something about your background, interests, hobbies, goals, and dreams helps people relate to you; plus, you may be surprised at what you have in common with others. I once had the opportunity at a break during a customer advisory board meeting to have a five-minute chat with the CEO. Our conversation eventually turned to a common interest in the use of technology for creating music. A few months later we met at another event, had a conversation about family and discovered that we lived near each other, and our daughters were both involved in swimming and close in age. This eventually led to an invitation to lunch.

Sharing your story also helps to dispel any preconceived (and possibly incorrect) notions of who you are. People are conditioned to group and categorize everything, and we do it unconsciously to others we don't know. By sharing your story, you become more relatable as an individual.

Value Diversity and the voices in your organization

A leader cannot be successful without the support of the people that they lead, and acknowledging this is vital. To take full advantage of the power and capability of the people on your team, you must take full advantage of their varied thought and abilities. I have found as the best way to take full advantage of diversity of thought is to ensure that psychological safety exists in all team meetings and one-on-ones, so team members can share thoughts and ideas on how we can enhance and improve what we do and how we operate. When you provide a fertile environment for people to share, great ideas begin to flow.

Sometimes new team or company members can feel hesitant to share ideas for improvements. In my organization, by giving everyone a safe space to share, new members of the team have identified options for process improvement. Some of these were processes that long-time team members had accepted as status quo. Encouraging newer members to speak up, sometimes a fresh perspective shines light on new opportunities.

Being genuinely interested in people's activities and passions outside of their current role helps provide insight. I had one team member in an infrastructure support role who was creative and interested in web design. When an opportunity arose to do some work in that area, she was excited to add this to her existing scope, plus it also gave her some additional visibility. Another member of my team was really great at building relationships with other teams. I noticed this and got him engaged with building stronger relationships with some of our external vendors, resulting in significantly enhancing collaboration and partnership with those vendors. As a leader, I have found that people sometimes see the unique value they bring to the table, but I can sometimes spot potential in them that they don't recognize.

It's not only important to value the diversity on my team but to use it as a strategic advantage. In doing so I make sure to provide support and encouragement particularly when team members are driving innovative change or expanding their scope into new areas.

***

Moving through my career, there were some things that I needed to learn, and some that I needed to *unlearn.* Valuing relationships and asking for help will get you through some of the toughest moments. Yet some of my most memorable lessons, and where I've learned the most, are my failures. All of that happened as soon I got outside of my comfort zone. That's when I truly began to grow.

Fred J. Love

Fred Love is Director of Unified Communications and Collaboration at Capital One. He has over 20 years of information technology experience with a strong and successful background in implementing and optimizing enterprise class unified communications, collaboration, and video solutions.

Leading is All About Asking

by Jeffrey Stovall

"The leader of the past was a person who knew how to tell. The leader of the future will be a person who knows how to ask." - Peter Drucker

"You must unlearn what you have learned." – Master Yoda

The thorniest change I've had to make in my managerial style in the past several years is "unlearning" how to ask. I was taught early in my academics that asking for help is not a sign of weakness. But asking can feel like a capitulation in a society that idolizes bold, independent thinkers. However, when one separates the heroic fantasy of executive work from its actual practice, one realizes that leading *is* asking.

Why *un*learning? Because asking in Western culture is often more than a request for information or action, it is an indicator of social status. We have all heard of asks that are not genuine inquiries. Some asks are debasements. Some asks are commands. Some asks are pleadings. The shape and form of asking reveals much about the power relationship between those in a dialog. So, to "unlearn asking" is mainly about unwinding some of the implied power structure of the act itself.

A mentor of mine, James Rodgers, is fond of saying that "executives do nothing"—an arresting statement that points to the essential role of a leader: to get things done *through others*. In this century, it is well-recognized that the leader as commander authoritarian is an outdated and less effective style. Leaders must make decisions, and leaders sometimes have to issue orders. But decisions and commands not rooted in trustful dialog and honest inquiry can readily backfire. So, what is a leader's purpose if leaders "do nothing" in an organization? Edgar H. Schein weighed in with a different take on what leaders do, in his book, *Organizational Culture and Leadership:*

> *"The only thing of real importance that leaders do is to create and manage culture. If you do not manage culture, it manages you, and you may not even be aware of the extent to which this is happening."*

Managing culture requires a feedback loop of inquiry and action. Leaders must ask, listen, and think before acting. I see three dimensions of "asking" which are crucial to sound leadership.

Leaders Ask to Understand

By now, many leaders are very familiar with Stephen Covey's *The Seven Habits of Highly Effective People*, especially number five: "Seek first to understand, then to be understood." In practice, gaining an understanding requires the leader to ask. The leader who assumes to have all of the information needed without asking is likely to run into blind spots rather quickly. All asks for understanding are not the same. Interrogation is a highly retrograde form of "asking," yet it is widely practiced. Asking without creating a sense of safety and collegiality is more likely to lead to misinformation than understanding. Edgar Schein writes in *Humble Inquiry: The Fine Art of Asking Instead of Telling*, of "the fine art of drawing someone out, of asking questions to which you do not already know the answer, of building a relationship based on curiosity and interest in the other person." To do so requires that a leader actively acknowledge both the limits of their own knowledge as well as the limits of their ability to succeed without the knowledge others possess.

In effect, the leader depends on those who are required to complete a task. In *Execution: The Discipline of Getting Things Done*, Larry Bossidy and Ram Charan note that "dialogue is the core of culture and the basic unit of (leaders') work." Bossidy and Charan write of "tough, incisive questions," whereas Shein calls for "humble inquiry." There is no getting around the fact that leaders need to establish trusted relationships that revolve around asking the right questions of the right people, staying engaged in getting the right answers.

The times in my life when I was blindsided by events that were likely foreseeable, it was almost always due to not asking enough questions.

In one job I took that proved to be a poor fit, I was flattered to take on a role of greater responsibility and title than I had previously held. I had a pre-formulated list of requirements for a job change, and I studiously went through the interview process, checking off my boxes for the job that I had idealized as my next best role. Throughout the interview process, the questions I asked tended to be of the softball variety because I had already decided that I wanted the position. Yet, the *unasked* questions were crucial for developing a deeper understanding of the circumstances that I would be entering. As a result, I missed some vital indicators that things were different from what I had assumed. Sometimes I ignored obvious warning signs due to my too-early conviction that the role was right for me. In the end, I did get almost all of my boxes checked. Several months later, I realized that *I did not have enough boxes*.

Had I engaged in more inquiry for understanding versus to fulfill my confirmation bias, I would have realized that the role would not work out as I had imagined.

Leaders Ask to Receive Help

Leaders must fully comprehend that they are dependent on the work of others to fulfill their objectives. A leader without willing followers is not much of a leader. Rensis Likert, decades ago, demonstrated that autocratic leadership styles were less productive than consultative or participative ones. Still, they persist for complex reasons. My distillation is that leaders commonly focus more on pride in perceived social status than on results. Without trust, asking from someone of perceived lower social status can feel like a demotion. Asking of those in a so-called higher social status can demonstrate that there is a status differential to be worried about; in the worst possible light, it can feel like begging. Yet, no leader succeeds alone, and waiting to receive help that is not requested is a gamble at best. Leaders must ask and ask in ways that open doors for future assistance. For a leader, asking of others is a way of building future relationships and getting current results.

In his book *Mastering the Challenges of Leading Change*, H. James Dallas says of those who ask:

> *An enthusiasm for asking questions suggests strong leadership potential. First, it shows that the questioners are always in learning mode and don't care who does the teaching. Second, asking questions is a practice that helps the people they interact with feel significant and respected. And third, it shows that they are comfortable with saying, "I don't know, but I will find out." This trait is essential to finding your way with confidence after you get lost.*

Personally, asking to receive help from more senior leaders has been a difficult change for me. Several times in my career, I found myself in roles without a clearly delineated mentor, sponsor, or advisor. Those have been some of the roughest times. I have generally been willing to ask for insights from peers and almost always willing to admit that I needed help from subordinates. Nonetheless, I was usually reluctant to ask for help from more senior executives because I believed I needed to act more as an equal to gain their confidence. In truth, that was a false choice. Asking for help was actually the way to unlock a broader set of insights I did not have access to in my position. Asking would not have been some broad-based admission of inferiority but an intellectually honest admission of limited perspective. We all have limited perspective. Emotionally intelligent leaders can accept this as a given while continually seeking to expand their areas of understanding.

This insight hit home for me a while ago as I sought the counsel of a founding member of ITSMF. I could not attend a session the member was to lead for a large group of CIOs, and I apologized, but also offered to buy dinner for the executive at an exclusive restaurant the night before. The executive agreed, and when we dined, she proved to be a fount of wisdom and insight into my current circumstance. In addition, she offered to facilitate introductions for me that I could not make on my own. The night was easily the best $300 I ever spent. In the afterglow of the discussion, I lamented that I had previously let so many years go by before reaching out to her. The access had been there all along; I simply had not asked.

Leaders Ask to Provide Help

If, as Schein wrote, the "thing of real importance that leaders do is to create and manage culture," then providing help is also part of the equation. Humble inquiry is an appropriate response to align a leader's perceptions of what is needed with that of the persons in need of assistance. Predictably, we equate offering help with altruism. That can be true, but more frequently in a business context, providing help to others can be the most direct means to accomplishing organizational or personal goals.

Offering help as a leader can be an accelerant to achieving results through others. It has the dual benefit of forging stronger relationships and can amplify a leader's proper role. Offering help is an assertive act of managing culture, enabling the ability to direct or model desired behaviors, and need not be viewed as manipulative or

Machiavellian; it is a principal mechanism of Bossidy and Charan's "dialogue."

However, in offering help to subordinates, understand that it can be delivered with an edge if the leader is not cautious. Sometimes, a proposal for "help" can be couched in a way that negates the safety of the receiver to accept it, coming off as a way for the person to declare their insufficiency to the task—a submission versus a helping hand. Typically, a solid pre-existing relationship and a clear view to prioritizing the welfare of the person being helped can offset those potential negatives.

As the new CIO for the City of Charlotte, I was having a tough time working with the head of a department who was very resistant to change. More specifically, change that was not entirely on his terms. I understood my assignment from the City CEO to lead change in an area that affected the department head, with whom I was at loggerheads after many meetings, conversations, and negotiation sessions. Finally, in a moment of honesty and vulnerability with the CEO, I confessed how challenging the situation was and why. He listened and asked, "How can I help?" I asked him to clarify for me what his goals were and the required timeline. The CEO agreed to do so and immediately restated his objectives for the project.

I thanked him for his time and support, resolving to meet him later along with the department head. A few days later, the department head's demeanor had changed distinctly. It turned out that the CEO was well aware of how prickly the department head could be and had a private conversation with him. The CEO had been looking for independent confirmation of his observations and was waiting to be asked to weigh in. Things did move along after that. The department head and I never became drinking buddies. Instead, we were able to hammer out a productive working relationship that achieved the goals the CEO had set.

Offering help to peers and more senior leaders can be a way of building bridges and exposure. As an executive advisor for a major software company, I regularly coach salespeople on positioning themselves with a CIO audience. I always advise them to offer to help their clients in a way that may have nothing to do with a sale directly and is not a quid pro quo. In my experience, CIOs are targeted by salespersons looking to "always be closing" rather than empathizing with the CIO's situation. Therefore, I coach salespeople to frequently ask, "How can I help you?"

This simple phrase opens a variety of doors for dialogue and is not premised on getting something directly in exchange for the assistance offered. On the contrary, a genuine offer to help another human being facing tough circumstances, even if it amounts to no material benefit to you in the near term, helps to cement trusted relationships that can have longer-term payoffs.

Marshall Goldsmith discusses the trade-offs between action, aspiration, and ambition in his book *The Earned Life*. A bias toward action or ambition can lead to looking at payoffs in the shorter term while missing the longer-term perspective of aspiration. Offers of help to others also conform to a timeline-based view of payoff. Is the offer simply about solving a problem in the here and now, or is it about achieving mutual goals? Perhaps it's about fulfilling who you see yourself to be and your purpose. Leaders ask themselves those questions before they offer help. Asking what you really want to accomplish is the ultimate act of self-leadership.

It's remarkable once you see it: the literature on leadership is replete with asking, listening, and following through—not commanding. I've learned that *leading is asking*. Now, what do you need to ask of yourself to be receptive to what you will hear?

Jeffrey W. Stovall

Jeff Stovall is an Industry Executive Director for Oracle Corporation, focusing on state and local government. Prior to joining Oracle, Jeff served as the Chief Information Officer for the City of Charlotte for ten years. Jeff was the 2022 Executive Member of the Year for ITSMF, was twice named a CIO of the Year by the North Carolina Technology Association and was recognized as one of Government Technology Magazine's "Top 25 Doers, Dreamers & Drivers."

Prioritizing: Who Works for Who?

by Charles Thomas

Honestly, I was unsure that leadership was the career path for me. For years, I was comfortable and did all the right things to not rock the boat. But leadership found me. Someone saw in me something that I did not see in myself.

Earlier in my career, a manager asked me what I wanted out of my job. My response was to make enough money to support my family, which seemed easy enough; that is what my father did for us. This manager shocked me by saying he saw more potential in me, that I was selling myself short and playing it safe, and that I should dream bigger.

Thus, my early journey into leadership was not born of ambition but through conversation. I have always had the desire to help and serve others. Having a family changed my goals in life. I wanted to provide a comfortable life for them, and that also served as an inflection point for me—I too wanted more out of life.

The moment that I decided I wanted to lead others occurred at my mother's funeral service when I saw a tall Caucasian man sitting near the rear of the church. He did not know I saw him, but I immediately recognized my manager. I was overcome with emotion: *He cared enough about me, his employee, to be here.* He was behaving like a shepherd to his flock. That day, I became dedicated to developing the same level of compassion and empathy for others as he had displayed.

Unbeknownst to me, my manager was preparing for a career change. In previous career conversations, I'd told him about being part of the student council in middle school, a squad leader in the U.S. Army, and president of the local chapter of my fraternity, and before leaving, he offered me the opportunity to be a team lead. After an interview with my

Director and VP, they let me know that I had the position. I had crossed over from being an individual contributor and looked forward to the challenges ahead to lead.

Now, over twenty years later, I have grown and hold the title of Director. And just as my first manager did, I am impacting lives. I will share with you some of my leadership principles developed over the years.

<u>Empathy</u>

Empathy is an essential part of my leadership style, as I strive to treat everyone according to their specific needs. This was one of the first lessons I learned, one of the best ways leaders can connect with their teams. Empathy builds trust through better engagement and open and honest communication.

When one of my teams was working on a significant project, one individual was not delivering on his action items, preventing the team from hitting its deadline. During a one-on-one, I could tell he was very anxious but felt comfortable enough to explain what was preventing him from completing work on time. I listened to his concerns and his challenges, and together we worked on a plan that would allow him to take care of his issues and also be successful helping the team achieve its goals. Together we were able to help reduce his stress and anxiety, ultimately creating a more positive work environment.

<u>I work for you.</u>

The strength of any organization depends on the strength of its employees. Creating a solid team means melding a diverse group that complements each other's strengths and weaknesses. I strive to foster a culture in which all members feel valued and empowered, with the goal of creating an environment where individuals feel comfortable enough speaking up, knowing their voices will be heard and taken seriously. Listening is essential, so I am a sounding board for any problems, concerns, or issues.

I let my teams know that my job is to support them, to help them grow, while we grow our business. I work *for you.* Deeds, not just words, embody this. By modeling the behaviors and values that I want to see in my teams, I lead by example. I demonstrate a strong work ethic, commitment to excellence, and a willingness to take responsibility for my actions. First, I must ensure that my employees have the tools and resources to do their jobs effectively and efficiently. While doing so, I aim to be positive and encouraging. Since a team usually reflects their leader's style, I constantly endeavor to bring out the best in everyone by trusting and respecting them, allowing them to do their jobs without undue interference.

Developing the team is an essential function of a leader. If team members end the year with the same level of skill and knowledge with which they began, their leader has not had a successful year. Influential leaders must invest in team members, providing opportunities for learning, growth, development, and advancement. An achievement I am very proud of is the number of leaders I have helped cultivate and grow.

One way I do this is to allow someone different to lead each monthly team call (where it aligns with their career goals). I also assign team members to collaborate with me on special projects to involve them in other areas of the business where it makes sense. In those conversations, they can often talk through a resolution, with my only participation being posing questions and scenarios.

Preparation

Being in the sales engineering world, I know the importance of preparation. Ensuring that systems are working, that demo environments are reserved and available, and scheduling dry runs can make a difference in sales. When teams are prepared and practice, we make better decisions and perform at our peak.

Recently, when my team was preparing for a competitive situation in demonstrating our solution to a large global customer, we knew their decision would be determined based on the outcome of the demo. The first vendor did a fantastic demonstration and was able to highlight their solutions flawlessly. We were next, but during our presentation, we lost our internet connection and access to our demo environments. However, because we were prepared, we could jump into a pre-recorded demo and continue our presentation. There were other issues once we were back online, but we recovered because of our preparation. We did not have the ideal demo, but we demonstrated to the customer how our tools could help them function and recover with minimal downtime and business disruption. Preparation allowed this leader to reduce stress and anxiety, increase productivity, and achieve desired results.

Prioritize Family

I always emphasize to team members that family and related obligations should be a top priority, investing time to build and maintain strong, healthy family relationships, and creating boundaries to protect that family time. That's not always easy, but it is essential to take time off, turn your phone off, and enjoy special moments.

I mentioned earlier that the team often resembles the leader's style and work ethic. As a father of three active girls with hectic lives, I often get pulled in many directions—often to events scheduled right after school while I am still working. I recall one very demanding boss who wanted projects done and completed in his time frame with no excuses or exceptions. This created a very toxic and unhealthy environment for me, and I often missed important family events. I had a candid conversation with my manager and established boundaries, producing a plan to enable me to succeed at work *and* participate in my children's activities. This lesson has allowed me to be more empathetic when my time came to lead.

Future leaders must take care of their physical and mental health, making time for healthcare appointments. Nothing else matters or can happen if you don't. Having—and eventually, being—a nurturing and supportive manager goes a long way to having a happy, healthy, non-toxic life. I now have open communication and an open-door policy about ensuring that family remains a priority for my team members.

Embrace Change

Leading through change is difficult, but if we can lean in and lead by example, focusing on the results, we can help team members and companies embrace new ideas without being distracted by all the outside noise and disruptions. Influential leaders are agile and adaptive while being honest and reassuring at the same time. I have been through multiple organizational and company changes via mergers and acquisitions during my corporate career; staying focused on the job, embracing the change, and staying positive have helped me and my teams through tough times.

Choices

Every day you are presented with choices. They can be small: what am I going to wear today? Or large: shall I make a job or career change? Everything we do is a choice, and those choices profoundly impact our lives and lead us down different paths, each with its own set of opportunities, challenges, consequences, and rewards.

When I was presented with the opportunity to lead, I made a choice that changed my career trajectory and life. I moved to a different state, started a family, and began a new adventure away from everything I knew. I transitioned from a small role as a team lead to moving upward toward executive leadership. At first, it was a tough decision, but I considered how it would align with my goals. Own your choices, and remember, doing nothing is also a choice.

Being Intentional

Intentionality means making a conscious and concerted effort to act with purpose and focus, doing what is essential to you—being present and in the moment, fully engaged in the activities that are important to you. This mindset requires self-awareness and an understanding of what motivates and drives you.

One way of being intentional with my teams is to ensure that during calls, I recognize individuals that have done something great personally, and following up with a direct email (or even a handwritten note) expressing appreciation for their extraordinary efforts and/or to celebrate significant accomplishments. This is a small but essential part of how leaders can show appreciation.

Being Accountable

I often ask my team how we can improve. More importantly, I ask for feedback on how *I* can improve as a leader.

Charles J. Thomas

Charles Thomas is a Solution Consulting Director at OpenText, a large cloud software company. He has over 20 years of experience leading high-performance teams, and he has held several technology and leadership roles across several organizations, working with sales, sales engineering, and sales operations.

Leader, Know Thyself

by LaKienya Williams

According to the Bureau of Labor Statistics, in 2022 more than 318,000 executive positions in corporate America, out of a total workforce size of 163 million. Obviously not everyone can be an executive leader; the title is granted to those who set direction and make decisions for the rest of the company. Not everyone with an executive title deserves it or is even good at it, but somehow, they must have done something right. Those who are truly respected and have the followership of others usually possess certain desirable qualities, including being a good communicator, trustworthy, relatable, decisive, good listener, knowledgeable, experienced, and having presence.

As I have grown in my career and moved up the ladder into management, inevitably I have also moved into leadership, although technically, leadership is not associated with any title or position. You can be a leader at any level and in any position. The desirable qualities are not exclusive to executive leaders but can be found in any emotionally intelligent, well-rounded person. Gaining experience and exhibiting leadership qualities will get the attention of those around you and increase your opportunities.

Yet even as this growth takes place, your idea or vision of yourself may not yet be aligned with what others see. Have you ever felt like you were not the leader others saw? Have you felt like you did not want to take center stage as you see other leaders do? Have you felt a sense of fear or heard a little voice that made you doubt yourself, thinking you couldn't do what other leaders do? That you are not yet comfortable enough, or get too nervous, or think others wouldn't listen?

I have had these feelings, and more. All the while knowing I am fully capable of being successful in this position and deserving of it—knowing I can lead others, because I'm already doing it, knowing I can handle the pressures and exposure because I am already doing it. So, how did I reconcile these two seemingly opposite voices within myself?

What helped me tremendously was working with an executive coach one-on-one. I had to work through those feelings of inadequacy which caused me to not have a voice, not want to be the center of attention, not want to "take up space." It never occurred to me that I was missing something so fundamental—that sense of knowing I am enough. The understanding that my authenticity, a masterpiece created by God, was supposed to be shared. And the same is true for everyone. You and your gifts should be shared. The crazy laugh. The inquisitive nature. The quiet thinker. The hugger. The planner. The visionary. All these traits and more should be brought with you to work. They are what make you, you.

<u>A story about voice</u>

I am a native Queens, New Yorker. Upon graduating high school, I attended Prairie View A&M University in Prairie View, Texas, traveling halfway across the country, from the city to the country. I did not visit beforehand, had never been to Texas, and knew no one there. What I had was a four-year scholarship and a determination to succeed in a challenging engineering discipline. Five years later, I graduated with a Bachelor of Science in Chemical Engineering, then started work for a consulting firm in Houston. I remember casual remarks from people at work about my directness in conversation. The remarks were not negative, but it was the first time I had heard this feedback. I really did not know what exactly was meant by "direct." But fast-forward a couple of years and feedback from my career counselors included that I wasn't assertive enough, was too quiet. Now, how did that happen? From being direct to not assertive enough? Somewhere along the way, I internalized those early comments about my directness and quieted my own voice. But why would I do that?

Some twenty years later, while working with my executive coach, reflecting on those early days in my career, I came to realize that while gaining an engineering degree was an achievement to be proud of, it was also a time I equated with failure and an internal loss of confidence. How so? Simply because I did not graduate with an engineering job offer. I thought I performed well in my job interviews, but they did not turn into offers. So, while I accepted a great offer from a consulting firm (and so did many others fellow engineering graduates at the college), that did not match my original plan or expectation. This one disappointment morphed into an internal voice that got increasingly louder: *you did not go to school to work in this field, you know less than everyone else, you will not measure up*. Michael C. Hyter, in his book *The Power of Choice: Embracing Efficacy to Drive Your Career*, describes this inner dialogue as a form of confidence-eroding self-talk.

That voice remained even while I gained the experience and knowledge that makes me successful in the IT industry today. The book *Humble Inquiry* by Edgar H. Schein and Peter A. Schein explains how our minds, through the data-gathering processes of the nervous system, determine how we react to things using our own biases and distortions to make sense of situations. And it happens within a split second. To answer my question from earlier: I went from direct to quiet so as not to bring attention to the fact that I did not belong (or so my mind said).

Was I not confident? I *was* confident (and I still am). I move through the day as confidently as the next person. But a wall prevented others from knowing the real me, and also prevented the real me from getting out. I was still rewarded with measures of success: promotions, great performance reviews, bonuses, trust and respect from peers and leaders. But deep inside was a mindset that I did not compare to those around me, which caused me to be quieter than I needed to be. This lack of confidence internally at the deepest levels was a huge weight to bear, even while those around me saw me as a leader on the rise.

Through one-on-one weekly coaching sessions, I discovered that my whole self was needed and that who I am is enough. As we explored concepts of value, voice, leadership, and management, I realized I already have everything I need. I did not need more degrees or more training. Just an acceptance of self and knowledge of my power. Once I had a grasp on that, feelings of inadequacy disappeared. The weight lifted. The unconscious distressful "energy" released. The facts are that I have been in this industry for many years holding several roles. I have the experience. I have the knowledge. I am respected for my knowledge. I know that I can do this, whatever this is. And if I do not know, I no longer fear not knowing but see those areas as opportunities to learn and expand my knowledge. What has risen within me is a belief and a confidence in myself from which I can launch into executive leadership.

Why do I tell this story? As a reminder that self-reflection should be a normal part of your routine so that you can address problem areas sooner rather than later. The importance of knowing yourself should not be overlooked! Doubting myself had become a habit because I did not know my value, my voice, the importance of *me*, and what I was bringing to the workplace. Do not wait twenty-plus years (like I did), allowing negative thoughts with no basis to fester and take hold. Leadership is a discipline involving skills which can be learned and honed for excellence. Practice makes perfect. The more you practice and become comfortable with what makes you *un*comfortable, the better you become at managing yourself and the uncomfortable.

My Core Leadership Tips

"People don't care how much you know until they know how much you care."
-Theodore Roosevelt

Get to know the people you work with as much as possible. This is my number one principle. People do the work for the organizations which results in success for corporate businesses, and they spend most of their waking hours at work. You want these hours to be productive. Creating a comfortable work culture supports productivity. How we relate to each other is of utmost importance. Everyone has family and other priorities outside the workplace that must be managed; everyone is juggling multiple demands. Understanding these other areas of importance for your employees, bonding over similarities and learning about differences, improves interactions and builds trust and respect. In my previous organization, I consistently received better scores than my peers when evaluated for employee feedback. Although I had a small global team and we were virtual, I was able to accomplish a lot through this team. When you are not physically located with your team members, it becomes even more important to spend time getting to know them.

Build a Personal Board of Directors

You need a support network. Navigating a career is not easy; you need the support and counsel of others—mentors, sponsors, coaches, sounding boards, confidence builders, and those who provide constructive feedback. Once you identify people for your board, put in the work to maintain these key relationships.

Challenge Ideas but Respect People

It is critical to challenge ideas in the workplace. This is how we progress and move forward. This is how we innovate. No challenge, no growth. Challenging ideas is not a personal challenge to any individual when done in a respectful manner.

Tell stories. It's how you relate to people.

Personal experiences are how people connect. I used to think I did not have any stories to tell, so I glossed over my experiences as if they held no value. This of course is not true, but that was the message I was telling myself. I now know the importance of reflection and acknowledging all that you may be feeling, thinking, and experiencing. Your experiences inform your future and your behaviors. Sharing them helps others who may be going through similar events. Use your stories as ways to impart learning, build relationships, and strengthen bonds.

You do not need to know everything.

No one knows everything. Even the most experienced leaders do not know the future and cannot predict business markets or strategic outcomes with 100 percent accuracy. Your value lies in your perspective and ability to guide your corporation forward. Executive leaders are bold enough to share perspectives and support ideas, without the goal of being right or wrong. I only just recently learned to be comfortable with this concept and not be unnerved sharing ideas. In corporate America, being intelligent will earn you some respect, but it is not absolute and does not guarantee anything. The corporate game is not won by intelligence or book smarts, but by how well you play the game.

Anything worth doing is worth doing badly. Yes, badly. Fail first/fail fast, a popular business philosophy, means you can learn fast and move forward. You must be willing to take some risks and accept failure. If you do not, you are not challenging yourself. You cannot grow if you are comfortable.

Bring your whole self to work. Trust that you are enough, use your voice, reflect on your experiences while finding ways to address negative thoughts as needed, and don't be afraid to fail. These principles lead to your growth and improvement making you a better leader by forming a solid foundation for executive leadership.

LaKienya S. Williams

LaKienya Williams is an experienced IT management professional with expertise in leading software development and operations teams. In her current role within HP's Data Analytics and Operations organization, she leads Data Privacy Consent Management, providing pan-HP support for various business units and commercial organizations.

III

Staying Vulnerable, Making Connections, Showing Up

The Better the Conversation, the Stronger the Connection

by Christopher Bowers

From early in my career, I asked myself, "What do our leaders *do* all day?" Given I was doing all of the work, I couldn't figure out how they were spending their time, much less what qualified them to be in their roles. This confusion, along with a lack of opportunities to lead myself, resulted in jaded feelings, some that I still wrestle with to this day. We used to joke that leaders did all the "*ings*" except for *do*ing. Now I know how true, and essential, that actually is. How, then, do leaders provide value to our enterprise? They seem to know very little about most workers' day-to-day yet set direction for where everyone is going.

It took a long time for me to change my perspective and stop being so arrogant. I started to surround myself with accomplished people. I started reading the books, publications, and articles they read. I began to have different conversations and finally, the aperture of my world view widened.

That's when my career started taking off. Senior leaders sought me for their leadership team. The most talented engineers began seeking my team out by name. We were delivering in a differentiated way. I believe I even became a measure for how well my peers were performing. We were winning awards. Something about me had transformed. What changed? I had discovered what it meant to truly embrace the mantle of leadership.

Most definitions of leadership sound impressive but are full of buzzwords and rhetoric. It wasn't until I was sitting in executive leadership training that I heard a much more palatable and simplistic explanation, from James Rodgers: "Leadership is the ability to sell an idea in such a way that people will follow you willingly."

How do I get people to follow me willingly?

For me, leadership begins with a deep knowledge of what is important to the leader as a person; this is the nucleus around which all decisions revolve. It determines what kind of company culture you will thrive in, how much time and energy you can spend investing in relationships, as well as your understanding of the business. Knowing what matters to you will drive your ability to influence, communicate, sell, find senior advocates, ask for help, and lead a diverse workforce.

Have you ever faced a decision in life where the answer was not obvious and there was no right or wrong answer? How did you ultimately decide? In these situations, I fall back to my core values: honesty, integrity, morality, quality. Understanding the other constraints in my life also helps: my family's needs, my support system, my relationships, my personality, my strengths, my weaknesses, and my goals.

This isn't a one-time process. You have to keep checking in on yourself because while values tend not to shift, life can change frequently and at times significantly. As my children grow their needs change; my spouse has a growing career; our financial situation shifts; a parent becomes ill; we make investments that need attention. Given all of that, I make sure to take time to check in consistently.

You can begin with this simple question I have often struggled to answer: "Tell me about yourself." To be an effective leader, it is essential to understand oneself. This enables leaders to recognize their strengths, values, and personality traits, and leverage them to build credibility and lead their organizations effectively. Self-awareness also involves understanding how one's values and guiding principles influence decision-making. Leaders with a clear understanding of their values can make decisions that align with their beliefs and principles.

A good example is landing my newest job. The catalyst for the change was achieving all of my goals in my previous role. I was getting bored, and I know myself. My bad habits appear as soon as I start feeling that twinge of boredom. I'm a natural problem solver and left unchecked I will create problems for myself to solve. Having been through this before and knowing what happens, I was able to take corrective action before I created a possible bad situation for myself and my team.

Meanwhile, I was getting feedback about experience gaps in my resume. I had spent so much time on workforce-facing applications that I needed more experience with large scale customer-facing applications to round out my experience. I eventually selected a job that was much different from what I had done before, the next step toward my long-term goals. Because I understood my needs, I was able to take my time, lean on my relationships, and select the right role once it became available. Those relationships include people that report to me, stakeholders, and other members of the organization. Demonstrating this competency is actually a measure of effectiveness in many corporate cultures.

I wonder how many self-styled geniuses, like me, have ever thought, as I once did: *The work will speak for itself. As long as I keep my head down, build the right things and achieve great milestones, my career will take care of itself. I don't want to tell people about my work because I don't want it to seem like I'm bragging. I don't need to do that, I'm not a brown noser.*

No! This is all wrong. Actually, part of it is right: The work does speak for itself. But guess what? It doesn't speak for you! If no one knows you or what your specific contributions are to that work, then how can they understand your value? How can they understand the value your team has brought?

Those with godfathers, get blessed.

Finding senior advocates is the pinnacle of relationship building. Have you ever noticed those special people who seem to get promoted every 12-18 months? There is someone in the organization who has taken a special interest in them, ensuring that they continue to get great opportunities to shine. In order to achieve your goals as a leader you must be able to identify key stakeholders who can support your aspirations, well-respected senior leaders who can provide guidance and support, and open doors.

I spent three years in a "stretch role" without getting promoted. I had incredibly high engagement scores, I had referred over 50 people to the organization, my key performance indicators (KPIs) were top of the heap, three of my former peers had been reporting to me for years. Still, I kept getting cryptic feedback; now that I'm aware, I understood it meant that I needed more sponsorship. There were naysayers on the executive leadership team who didn't know me and were willing to explain away all my successes. I needed to take more time and focus on gaining their trust.

Things did not change for me until I decided to make a great relationship with our new division CIO. I did my homework. He was new to the company, struggling to adjust to our culture, and an engineer at heart. I decided to eschew my normal approach of ignoring leaders thinking they will see my work and assume I am great. Instead, I asked him for time, and we hit it off very quickly. I tailored my overview of my team as if I was speaking to one of my engineers. Very cut and dry answers, no flowery language, no buzz words. He loved it and told me that I was the first person to truly be able to answer his questions with clarity—and immediately booked more time with me. We grew to be friends. I became his bridge to our company's culture. He called me when things were not going well to get perspective. He offered me differentiated experiences.

I got promoted within three months. The best part about this situation is that I did so much less "hard work" during this time as compared to all of the toiling I had done in the past. It felt wrong but it taught me the biggest lesson of all: build strong relationships. You'll stand back one day and realize everything has changed for the better.

Christopher M. Bowers

Chris Bowers is a game-changing software executive with nearly 20 years of experience in leadership, software engineering, and architecture. He has held several successful cloud and technology transformation roles in both corporate and startup settings, spanning across healthcare, fintech, and defense industries.

Leading with Transparency and Vulnerability, as a Space Creator

by Shawna Brunson

A title does not make one a leader. People with titles that indicate leadership (manager, director, etc.) may not be leaders at all. And those without leadership titles may in fact be leaders. Leadership exists at all levels of an organization. True leaders are fair, respectful, and take accountability for the decisions they and their teams make. How do I know? Experiences!

Throughout my career, I have been fortunate to work with some amazing leaders who in turn have inspired my own leadership style and helped me understand the type of leaders with whom I want to work. Now, I pride myself in being a leader who is transparent, vulnerable, and who wants to provide a space for people to learn and grow new skills.

<u>Transparency</u>

I learned early that transparency requires work and diplomacy—and a sense of timing. A good leader knows enough about their audience and timing to determine if transparency in that moment is of value to all involved.

I learned this from personal experience around displacement of team members—something which every leader at one time or another a leader will face. When displacement happens at a firm the size of Wells Fargo, team members start asking one another and their leaders if their name is on the list. In 2008, at the height of the mortgage crisis, the firm was letting a lot of people go, something that was in the news. This was my first time as a leader being accountable for executing displacements among my team.

I was meeting with my team of about 15 junior level associates and one asked how our area might be affected. My manager had said this would happen but still, my heart sank, and I wasn't sure how to respond. I took a deep breath and thought about something my manager had once said: "You may ask me something that I know the answer to, and I will tell you so, but I may also tell you I can't share those details."

This was happening at a time of conference calls without video, so I couldn't see the reactions. I responded by saying something like this: "Our area is considered administrative/overhead. What this means is that as the business adjusts based on the market, we adjust. I can't tell you when and I can't tell you who, but we all need to understand our role in the organization and understand the commitments to the shareholder which sometimes includes reducing costs, and the highest cost is our labor."

Even today, years later, I still remember the silence and fear that seemed to come across the phone line following that statement. I questioned my own actions at that time but would learn later that those words had an impact on two people in my team, spurring them to immediate action. One team member, who was part of the planned displacement, talked with her family about her desire to return to school and finish her degree so she would be more marketable in the future. By already starting that discussion with her family, she had concrete plans in place for using part of her package. Another team member, who was actually not displaced, took the opportunity to cross-train on other capabilities to make herself more marketable within the firm. I had been as transparent as it was possible for me to be at the time, sharing what I could with those affected by the information. As a result of my words and the idea of "adjusting as the business adjusts" these two women were able to make plans and be ready.

Vulnerability

Vulnerability, or as I like to think of it, being human, is an essential quality for a leader, and yet we also must acknowledge that in some industries and in some circles being vulnerable is not welcomed. I am a leader who wears her heart on her sleeve. In my 26 years at my firm, I don't know if you can find a current or past team member who hasn't seen or heard of me crying over something sad, funny, or inspirational. I believe this human vulnerability requires knowing each other on a personal level. It doesn't have to mean engaging in social activities, but rather taking the time to talk to team members about something other than work so that you can start to know each other as people.

To that end, I begin every one-on-one call and small group call by asking about family and friends, taking the first five minutes to converse about something other than work. Doing this allows you to get to know someone and also lets you notice when someone is going through something and may need a little grace from you and others.

This approach to getting to know people came in handy with a manager who worked for me about ten years ago, a leader who was always on time with deliverables and engaged in roundtable discussions with me and his peers. He had worked for me for about two years when I noticed his work slipping. He was late with deliverables; his team began complaining about his lack of engagement; I would have to remind him of things more than once. These behaviors were uncharacteristic. As his performance continued to decline, I asked him if he had taken on more than he should? Did he have outside commitments affecting his ability to deliver? And was he enjoying his job versus being merely content? Not enjoying one's job can cause work to slip.

On the fifth or sixth time we talked I could hear in his voice that something personal was going on. I knew that he and his wife had been married for over 30 years and that he had a son in his early twenties. I knew that his family had been going through some stuff with his son but based on recent conversations, he and his wife were still happily married. Then, with a hesitant and cracking voice, he confided, "I think I may have dementia. I just can't remember things from one day to the next." This was not what I was expecting. He explained that he hadn't seen a doctor yet, but he thought it was the early onset of dementia, which ran in his family. When he finally did see a doctor, they told him it wasn't dementia. He was confused. How could he be forgetting so many things if he didn't have anything wrong with him? What he didn't realize was that the doctors weren't saying he didn't have anything wrong with him, what they were saying was *it wasn't dementia.*

Fast forward a few more weeks, a few more doctor visits and discussions with his wife, and he decided to see a therapist. He would later learn that as a result of some early childhood trauma, he was experiencing this current memory loss in his fifties. While he was grateful to know the cause, he was not looking forward to unpacking those issues. I encouraged him to take advantage of the benefits he had paid into all these years and arrange a leave of absence. His fear of people only remembering the last six months of his work was something that caused him to stay working longer than he should have. He would eventually take that leave of absence and while on leave wrote me a detailed email explaining the journey he was on. He thanked me for recognizing that what was going on was bigger than the job or the firm and encouraging him to practice self-care. He retired shortly after but still sends updates about his retired life every now and then.

Creating Space

Leaders need to provide people a space to grow and learn. I used to say, "I felt lucky that so many people gave me different opportunities in my career that helped get me where I am today." I said it one time and another leader responded that he had always assumed that the "door was opened based on something I had already shown them I could do." He reminded me that no leader wants to give someone an opportunity unless they know that person is capable of delivering. But if you are in a position to give someone you have faith in an opportunity to grow and get greater visibility across the organization, that action will pay back two-fold throughout your career.

One of the ways I help create space for others to grow is by allowing people to use my area as a jumping off point. I have done this for some people early in their careers, and for others who have been with the firm twenty-plus years. As I move throughout the organization, I keep connection to my prior teams. This may be just a ping every now and then or a one-on-one meet-up. Part of that is me letting them know someone is here to help them as they navigate the firm, as others did for me. It is also about my desire to make things better for those who come after me.

I once had a side conversation with a peer who told me about someone on her team with an interest in doing technical design, which at the time, was only executed in my space. She told me how talented this team member was and that she would hate to lose him, but he was starting to look at other opportunities. He was doing production support, a 24/7 job and was getting burnt out. Since I had done both production support and technical design, I understood that they were vastly different and let her know that if she wanted to allow him time to shadow my team and maybe get some hands-on technical design experience, I was happy to provide the opportunity.

A few months later he started shadowing my team and within six months he was getting hands-on experience. I wanted him to test the water and see if the steady more predictable work and slower pace in design would be fulfilling enough to move out of a space where every day is different. If he was someone who enjoyed an ever-changing environment, then design may not fit his needs as he would work on the same design for extended periods until it was ready to hand off to the engineers. At the end of the cross-training window, I met with him to see how the experience was and this helped him decide where he wanted to go next with his career. He was in his early twenties. Not only did he enjoy the design role, but he also enjoyed my space so much he wanted to hold out for an opening in my team. Justin continued to work in my space off and on until he officially joined us six months later. I have done this same cross-training with several other team members over the years all with different outcomes. Whether someone joins my team through a cross-training opportunity or via the traditional route, I tell everyone if at any time they find this isn't the right place for them, I will help them find something else.

While I value transparency, vulnerability, and being a space creator, I am not a perfect leader. Taking the time however, to re-consider these qualities has given me an opportunity to remind myself of the leadership qualities I value, work to embody, and which form the core of my philosophy about leadership.

Shawna S. Brunson

Shawna Brunson is a Senior Vice President and leads Consumer and Small Business Banking Technology, Small Business Journey. Shawna has responsibility for all technology channels used by Wells Fargo customers who own small businesses, ensuring easy access, low friction interactions, solutions to manage cash flow, borrowing money and leveraging technology to grow.

Career Arc Constructed by Connecting

by Frankie Hill

"Let me talk to you about leadership."
– Willie Martin, Vice President & General Manager, DuPont (retired)

A senior executive at DuPont taught me a vivid lesson about leadership. Though his tutelage occurred outside of a formal training scenario, the impact resounded magnificently. This individual impressed me in manifold ways, particularly his capacity to provoke devotion. He garnered admiration from front line operators, career professionals, and senior managers. Across multiple constituencies, he commanded respect. He problem-solved with virtuosity. When there was a chemical release at a local plant, he was the one to speak to the media. He got tapped when disgruntled employees planned a march around corporate headquarters. Exuding charisma, he entered rooms and shifted the focus, though executives who were more senior than him were present. His merits meant immense demands on his time, yet he made himself available to speak to a young project manager in Information Technology, a task outside his core duties. In performing that kindness, he illustrated a profound insight for me: relationships matter.

People often speak about networking. While this activity opens doors, ideally it's only the start of a more meaningful process of connection. Interacting with my mentor clarified this. If you asked him for fifteen minutes, he would spend an hour and a half with you. Weaving between stories of a personal nature and inquiries about my aspirations, he imprinted upon me very early on what an executive walked like, dressed like, talked like. He spoke in measured tones using clear and precise language. He tried to poach me for roles in his operations and manufacturing shop.

Through these activities, he expressed belief, provided guidance, and inspired me. He understood an essential truth—that leaders must touch people. They must engage with folks all across the organization, connecting on a personal level and establishing common ground. From the moment I started in corporate America, I witnessed the power of such efforts.

My DuPont origin story commences with graduation from high school and receipt of a corporate scholarship. Given the company's base of operations, I expected that my first internship would be in Delaware, and indeed that was what materialized during the summer of 1993, working in the corporate data center. Because I didn't have a car, I made the journey from my local lodging to the suburban IT offices by taxi. I had calculated the costs for this travel arrangement for the entire summer and planned my budget. After four days of executing this scheme, I arrived at work proud of my independence and initiative. But then, my problem-solving prowess got overruled when a colleague, Miss Linda, just returning from vacation, learned I had been taxiing to the building. She immediately arranged for ten colleagues, including herself, to provide rides. I learned then the power of having a support network. You can go it alone, but that strategy carries a high cost. By existing amid individuals who promote your growth, you not only thrive but also experience fulfillment.

Connection provides benefits through overt acts of goodwill. As my supervisor on that first internship taught me, connection also allows you to expand your circle of teachers. He provided a list of no less than twenty-five folks that he advised I needed to meet over the course of the summer. From that list, I discovered a web of collaborators who have nurtured my career. I appreciated the explicit help, and over time, I've noted an unintended revelation. My supervisor's advice reflected DuPont's status as a "relationship company." I later confirmed that this attribute is not peculiar to DuPont but applies to many organizations within corporate America generally.

Though folks eventually will return your cold call or answer your email, they often reach out first to someone in their network to determine your provenance. They seek a shorthand way of gleaning your inquiry's importance and your character. Performing this vetting process, they acquire a short-term answer and a long-term outlook. They learn about your personality traits and career arc. They listen, seeking points of connection and planting the seeds of trust. As these seeds are cultivated over time, a good colleague can move from co-worker to friend to someone you consider family. Such tightknit interactions have positive consequences. For example, when engaging partners from external vendors, I don't want my first time to be prompted by a crisis. Instead, I want it to build on a foundation of familiarity. Though I still ask for immediate attention when confronting a problem, I do so in the context of a protracted web of interactions. My call for help *asap* might sit alongside a quip connected to our regular trip to happy hour. Such texturing enriches shared endeavors.

A leader takes these lessons of building relationships, forming a support network, and learning from others into every subsequent role, pouring connection into the lives and careers of others via meaningful coaching, mentoring, and sponsoring. Recognizing this truth, leaders should strive to be selfless. As they understand the wants, needs, and aspirations of others, they can then tailor their assistance. On some days, perhaps a gentle push. Other moments, a thoughtful introduction, a useful suggestion, or proximate and contextual feedback. Ideally, these lessons are timely and specific and communicated stressing the existence of options so that based on one's advice, the listener will ultimately decide the appropriate course of action.

My approach to coaching and mentoring hinges on earnest conversation. When I meet you, I assess who you are and how I can connect you with simpatico individuals who can aid you. I try to engage, manifesting curiosity about your path, your background, your aspirations. This is an opportunity to listen to their story, tell a bit of my own story, find a common thread of connection, and help me to learn and improve.

Once a leader absorbs the lessons of building rich human relationships, they must execute. The simple truth is that you need others to get stuff done. During an episode in my career, I saw the complicated contours of this truth. Individuals in the IT global leadership team were discussing my capabilities, and some contributors, while acknowledging my track record of getting things done, raised questions about my assertiveness. They referenced my quiet and measured comportment and wondered whether, when the time came, I could muster an edge. After listening to this exchange, a sponsor who had interacted with me extensively corrected the inaccuracy of such observations. She supplied examples of my ability to provide forceful leadership. Because she knew me both at the office and beyond, she could offer vital perspective. Her intervention drove home the tight tie between human capital and successful outcomes.

I grew up around individuals who believed in "respectability politics." According to their example, dressing appropriately, working resourcefully, and minding your business would secure your place in the workforce. I learned that this outlook would keep you gainfully employed; however, it would not propel you into leadership positions. While aspects of my character matched with prescriptions of cautious reserve, I also possessed a more gregarious side. I soon realized the necessity of bringing more of myself to work.

Because my sponsor had seen an extroverted version of me, she could advocate on my behalf, creating opportunities that advanced my career. I had earned her vouchsafing by inviting her and others to know and understand the full range of my capabilities. This posture requires intentional practice. While toiling away in silence and obscurity may suit your personality, exhibiting more of yourself allows benefactors to expend their personal capital to elevate you when the need arises. Mentors can only act on what they know. When my sponsor endorsed my talents, she set in motion a series of assignments that augmented my capability and increased my competence. Nowhere are these changes more evident than in my communication skills.

Inevitably, leaders must deliver tough messages. While your initial strategy may include gentle chiding and cajoling, sometimes colleagues don't take the hint. Then, in the vernacular sense, you may have to "talk bad" to them. This is not about being abusive; rather, being as clear and succinct as possible. In the end, you need your audience to understand. Since they didn't get the hint and the memo went right over their heads, you abandon subtle tactics and move to maximum candor. You provide examples that corroborate your message. As you explain the situation, you express that if the behavior or actions continue, their employment could be at risk. These conversations are never easy, but when you have an established relationship with the co-worker, your message lands more effectively. They can believe that your critique is constructive and originates from a desire for shared success. When delivering such feedback, I find it useful to provide as much context as possible.

Where confidentiality or non-public information is not at risk, I detail my thought process and clarify the why for the conversation. This disclosure empowers my employees, equipping them with the understanding to make confident decisions moving forward. A manager who later became a sponsor once told me that at a certain point, your influence is about how much you can do through others. You maximize your capacity to effect change by empowering others. I enjoy orchestrating, bringing others along while supporting them in the background. By engendering trust, listening, and clearing obstacles, I can pursue my natural leadership style. I am more effective because I can lead organically without obsessing over who takes credit for progress. To be comfortable in this lane, I have to accept and sometimes share my default assumptions.

When seeking to transform organizations, I have found that it is better to try every approach that occurs to you. *Did I speak to everyone involved? Did I enlist the help and assistance of vendors and strategic suppliers? Did I take a bottom-up, top-down, middle-out approach? Did I explore every angle to increase the probability of success?* I have to play chess when others play checkers. If I am not thinking beyond the next move, I cannot stay ahead of my company's competitors. In the same way that politics is not a spectator sport, success in corporate America is not for the faint of heart. Excuses shorten careers. If you want to sleep soundly, do everything in your power to achieve your goal. Then and only then, can you contentedly await the resolution.

Corporate environments can sometimes breed secrecy. While such concealment may be purposeful and strategic, sometimes it is unintended and disruptive. What may seem clear and simple to you can often confound others. If indeed there is method to the madness, then pause to explain it. There is an old African proverb which paraphrased, speaks of *going fast alone but going far together*. If you measure your impact through the number of folks that you touch and inspire along the way, then taking individuals into your confidence is imperative. This requires prioritizing any employee in your organization, finding and making time to meet, to listen, to offer advice. My mentor, my captain, my best friend, my ace, shared that he poured into me for I was the only one worthy. I want my own legacy to be legion. So, I commit to do everything in my power for whomever I encounter. Make no mistake, it's a takeover, an attempt to leave a memory of service, devotion, excellence, and achievement.

The simple equation of expanding your scope, increasing your visibility, and seeking multiple individuals to support you provides a roadmap to success in any endeavor. This approach will increase your influence. No longer will you ask what you can do to stand out. No longer will you question why your efforts have not been rewarded. Your focus will shift to uplifting others. Selflessness liberates your thinking. To paraphrase the old gospel song, *If can touch the life of one, then my living will not have been in vain.*

You alone bear responsibility for pursuing your success or failure. When a manager told me that if I applied for a new job, he would bar me from resuming my old position, I ignored his warning. My bid was unsuccessful; however, I moved forward with clarity about my intentions. My failed candidacy brought me to the attention of a sponsor who provided a different job with essential experience. Because of this interlude, I purchased my first home and eventually secured an international assignment. None of these accomplishments would have occurred had I remained paralyzed. By being brave, I saw the difference between actions driven by determination and those sparked by emotionalism.

The foundation of success in corporate America is relationships. Once you embrace your aspirations, meaningful connections with others facilitate your effectiveness. You will encounter challenges in the form of issues with colleagues or structural problems in an organization; to meet them, you must create your own support group. The Bible teaches that we should do unto others as we would have them do unto us. When this maxim is applied to teams, it outlines a particular path to individual and collective ascension. Walk boldly.

Renarde Hill

F. Renarde (Frankie) Hill is a senior IT leader with a demonstrated track record of leading complex enterprise technology transformations. He currently leads a global organization delivering data, analytics, and digital solutions for DuPont.

Owning a Mistake Made me a Better Leader

by Michelle Ladkin

The question, "What does leadership means to me?" swirled in my mind after our first ITSMF Executive Academy module. The more we explored, the more alive I felt as a leader, actively focusing on leaders' messages, actions, and behaviors. Little did I know my leadership vernacular would improve and become a game changer for me—especially the idea that leaders lead by example, own their mistakes, and take feedback. These are now key components of my leadership style. Yet if I think about it, that all started much longer ago.

My parents, Ethel Mae Jenkins, and the late Arthur Jenkins, always said I would be a good leader. They emphasized to me and my siblings, "You can be and do what you want, if you put the work into it and believe in yourself." This belief became a part of my leadership approach, though early on I did not realize these leadership elements were fundamentally being folded into my daily life. During my junior and senior high school years, I was captain of a sports team, and while my main objective was to win games, I unknowingly also influenced team members. If one team member was having a challenge with any drill, I provided what was necessary to improve their skills. Little did I know, during this process we were creating a trusting relationship and other team members began to voluntarily follow my lead. While we ultimately won many games, the greatest achievement was that we also had fun and I'd helped elevate their skills, the team becoming better than what we were. As leaders, we should consistently work on improving ourselves, while also helping our team achieve the desired objectives.

"I have short-term goals—to get better every day and to help my teammates every day—but my only ultimate goal is to win an NBA championship."
- LeBron James

During my early years becoming a leader, I learned an eye-opening lesson, stemming from a mistake. The company was going through a resource reduction, and throughout this process I failed to communicate to my team the nature of the resource reduction. I had many new employees on my team and over the years, my area of responsibility was mostly untouched. Given the critical nature of our work, I assumed my team was fully aware we would get through this reduction process with minimal impact. Little did I know, many negative narratives were already circulating among my team—all due to not hearing any information from me. By the time I realized what was happening, I had lost key team members. The mistake of not communicating early and often came at a tremendous cost to the team, company, and me.

Finally, I called an urgent meeting with the team, humbly owned up to my lack of communication and vowed to be more transparent going forward. I knew team members would probably not speak up during the meeting, so I needed a strategy to hear their concerns and gather feedback. Prior to the group meeting, I met with one of the most unhappy team members to ask for assistance. Wow! Did I get an earful. Since I wanted to lead my team more effectively by improving my communication, I was prepared to rectify my mistake and the feedback was what I needed to help me to course correct. I listened intently and asked Mary to help me during the meeting by voicing her concerns and providing feedback, suggesting ways of improving my communication with the team. I needed someone to break the silence during the meeting and Mary was able to do this while also encouraging others to provide feedback.

Later, I incorporated a few suggestions from the team and made other chances to my everyday leadership, including a more inviting open-door policy; holding more frequent one-on-ones; establishing during team meetings a special time on the agenda for team members to voice their concerns and/or bring up topics they wanted to know more about. The walk-on topics discussion was a little uncomfortable for me in the beginning because I had no idea what might come up, but over time this became easier.

Through this journey, relationships with my team members shifted more to the positive as I grew more open-minded, developing a double dose of trust. In return, the team became more understanding when I said I was not able to discuss some of the topics due to corporate priorities. People want to see their leaders as human, someone they can relate to. Simple as that. This was not a new concept, but something I understood instinctively from my teachers and coaches when I was a high school student. But the lessons of my communication failure around the reduction topic created a major awareness that stayed with me throughout my career; a lesson that I now faithfully discuss with all my mentees as well my daughter, who has heard this story so many times that she now cites it as a learning lesson in her discussion with others.

All leaders make mistakes. How you respond to the mistake is what makes you a good leader, one who is accountable, takes ownership of their mistake, and moves forward learning from their mistake.

Michele J. Ladkin

Michelle Ladkin is Sr. Manager, Applications Development at American Airlines. She is a long-time member of American with over 37 years of experience in a variety of roles.

In the Presence of Leadership

by Anundrei Oliver

What is executive presence?

For one, it is the difference between the way in which an individual contributor and an executive leader approach a situation—a specific state of mind. When that state of mind is consistently and sharply focused—what I refer to as being "in presence"—that executive will have a positive, significant impact on an organization.

Accepting the Call

Those with executive presence are keenly aware of the broader objectives of their organization, and the need to achieve buy-in from stakeholders, to answer the broader questions, "Why are we doing this?" and "What's in it for me?" Someone with executive presence will focus on the broader team and aligning the skill sets that have the best opportunity to move from a "solving to" mindset to a strategy of "solving through" the work of others.

Executive presence requires humility, so evident in great leaders who are not only concerned with the outcome, but equally aware of impact and how their strategies reflect on the organization. The realization of that impact drives the executive leader, so that they are what I call "in presence."

Whether you lead one or one thousand, ultimately people seek out stability and leadership in tough times. I have had a front row seat to executive presence the past four years, observing leaders deal with some of life's toughest situations. I have seen candor, vulnerability, and a refocus on faith and wellbeing for the answers that are ailing our world. This has shown me that executive presence requires a greater perspective on mind, body, and philosophy, and requires continually reinventing ourselves.

It is not a secret that the life cycle for executives is short because of the demands of the role. There is a *secure your mask first before helping others* mantra that exists in leadership. Being "in presence" is practicing and actively promoting better health in your organization so that you are physically and mentally able to tackle challenges. It is making personal connections, and creating bonds and ties with the people that you lead. That connection gives clarity to your thought and allows others to understand you as a person, adopting philosophies that align with the organization. The way that you appear visually and the way that you engage in conversation makes an impact, putting you in presence so you can read the room, measure your delivery, and remain comfortable in silence.

Thought Leadership

Few decisions in leadership are made in a vacuum, nor do most decisions benefit from delayed action. By being in presence, you facilitate learning, arrive at conclusions, and deliver outcomes by creating an environment of thoughtful collaboration, by asking the questions that drive meaningful interactions. In presence isn't as much about the answer as it is about the journey; facilitating dialogue invests the team in the outcome and creates lasting bonds between team members and leaders and the organization.

Perspective as a Leader

It's often said that great leaders are not born, they come to be. I did not begin this journey with aspirations to lead but I've learned that I do have an ability to see through the conversation, and a desire to solve problems, which has created success for me. I was not even aware of the fulfillment I would experience in moving to serving people and the challenges we face as an organization. I've learned to be in presence, because that's when I have the ability to influence and impact others in ways that in the past were unknown to me.

My time in the ITSMF Executive Academy— the opportunity to mentor and be mentored, the shared education—was life changing for me, giving me the opportunity to perfect traits that will be pivotal going forward. "Leaders don't do anything" Dr. James Rodgers repeatedly told us. And while I believe that in the physical state, I believe the impact we have as leaders takes the "do" to a different place, especially when viewed through the lens of executive presence.

In presence, we can fully engage and connect to the person on the other end of that conversation with the intent of solving the problem and understanding the journey. Presence demands active listening and forgoing the inclination to solve the world's issues, instead listening for the world to be heard. An executive with presence can ask or answer complicated questions, address the elephant in the room, or stay comfortable in the silence in troubling times when we must challenge the ways of conventional thinking. As executives, we begin to look at the way that things have been done by our predecessors to understand if a current situation requires a different train of thought. It's that wisdom and that courage that transforms an executive's presence and separates the most effective leaders from all the others.

Talent Management and Social Impact

As an individual contributor, you are tasked with understanding the trends, fads, and the opportunities in only a given space. An executive who is in presence creates a culture of learning and a sense of urgency so that even as trends and fads present themselves, they see the way through and can drive individuals toward greatness, despite ever-changing technology that makes it tough to stay current. Leaders with a sharp presence create opportunities and support individuals even in turbulent times.

Frequently, multinational organizations split up into groups or areas of expertise inside of regions, and where large enterprise customers exist, often segment further to provide a hyper-focus for given technologies. This provides excellent client-to-customer interaction but can create environments that make it difficult to connect employees to corporate culture, especially post-Covid. An executive leader needs to bridge possible schisms.

Today's workplace is unlike anything that we have seen before, requiring a different leader whose presence is capable of tackling topics like diversity and inclusion, reassessing traditional ways of team building and problem solving. Unlike in the past when leaders often sought to project an air of impenetrability, today there is comfort in allowing yourself to be vulnerable to a broader audience. Being vulnerable in presence means to do so without regard to how that moment might play out. Yes, sometimes leaders must be the voice of reason, but just as often, today's leaders bring executive presence by acknowledging openness to others' input. These times when we change or flip the narrative, or when we must go against the grain when the values and principles we believe in come into question, whether they be times of turmoil and misinformation, allow us to show up in ways that will reflect the ideals of our executive leadership.

I have learned through my time in the Executive Academy that the power and connection created in relationships can be life changing, often in ways that at the moment are not always clear. There's never a shortage of people creating opportunities for happy hours or various after-work opportunities, but the emphasis needs to remain focused on relationships and relationship building. In the EA, we talked about family, life, and the trials and tribulations of a workforce in which you may be the "only" in a world that oftentimes seems like it's not suited for us. The relationships forged in that environment can be enormously beneficial if are candid and vulnerable with each other, so we can grow as individuals and create real bonds.

One thing that I personally did while in the EA was to be intentional about relationships, to go beyond day-to-day conversations and seek out cohort members in an attempt to create personal connections. Like anything that we want to grow, we need to feed relationships to ensure they will continue, to sustain and further nurture bonds.

Anundrei L. Oliver

Anundrei Oliver is Director, Customer Experience Product Management at Cisco. In this role, he leads teams through development of rich customer experiences across an industry-leading $4B security business with 2,000 team members.

IV

Whatever Way that Works:

Personal Inroads to Leadership

Leadership: What It's Not

by Shari Bradford

The year I was born, Thurgood Marshall was confirmed as the first African American Justice of the United States Supreme Court. Aretha Franklin's song "Respect" rocketed to the top of the music charts and Muriel Siebert became the first woman to own a seat on the New York Stock Exchange. That summer, race riots erupted across the country while 10,000 "hippies" converged on San Francisco's Haight Ashbury neighborhood. The Vietnam war raged on while the Apollo space program promised to land a man on the moon and the Mariner 5 probe flew past Venus. It was the dawn of the budding feminist movement, when the promises of space flight and Black Supreme Court Justices made it seem as though many things were possible for a Black girl growing up in America.

Across the San Francisco Bay, in Oakland, California, I grew up in a working-class neighborhood, born to a teenage mother and a father who sacrificed a college education to provide for his family. As he ultimately came to raise us as a single father, he taught my older sister and me to be confident, articulate, and unshakeable. Growing up, I watched my father work hard every day for others and occasionally at entrepreneurial efforts to meet the bare necessities in life. We were blessed with a beautiful, inherited home, and never wanted for our basic needs. But if I wanted something that my single father could not afford—and I wanted many things—I needed to earn the money myself. For me, it was clear that the way out of tough circumstances was to work hard, go to college, and get a high-paying job.

As I excelled at academics and got myself through college, my goals were financial freedom and to be the boss. Leadership at that time to me meant presenting a strong front, expressing my opinions, and telling others what to do. Over the course of my career, I have refined my thinking about what it means to be a leader. Even so, imagine my surprise to learn during ITSMF's Executive Academy that much of what I thought about being a leader was simply untrue.

Leadership is Not Doing

Growing up, I was no stranger to hard work. I knew my single father had sacrificed his own college education to raise two girls, working hard for others in low-paying jobs. So, I believed that the secret to success would be to keep my head down, follow my father's example of hard work, doing my job well, and letting my accomplishments speak for me. Then surely, my managers will recognize my work with a promotion. Many employees who aspire to senior positions within their company share this same belief: that hard work will advance you. This is perhaps one of the most sinister beliefs because it's partially true.

Hard work, productivity, and competence *will* always be valued in the corporate world. Employees who consistently produce, accomplishing the job that they are hired to do are the desire of every manager. A good work ethic is not something that you can teach. And although you may be able to teach competence, you can only add so much to a person's innate abilities. An employee who is both competent and has a strong work ethic is valued. Certainly, advancement should follow for such individuals. Yet I learned that doing a great job will only get you so far. Research and anecdotal evidence show that many competent individuals find themselves stalled at a certain level, often around Director or Senior Director.

Instead of a meritocracy, corporate America is more complex, influenced by a variety of factors beyond pure merit. Advancing past middle management and into the executive ranks requires more than simply doing a good job; having a strong track record of accomplishment is a given, and surely vision, strategic thinking, and likeability are also important. But to increase one's chances of success requires focusing on developing additional personal qualities and relationships.

Executives tend to choose employees to join their ranks who are likeable and whom they feel they can trust. Sometimes, what one lacks in competence can be compensated for in likeability and trust. That trust is based on an executive being able to see you in an executive role—you look the part, speak the part, and perhaps you participate in the same activities as others in more senior roles. Corporate culture demands assimilation—how well you fit in with the existing norms, values, and practices of the organization: things like adapting to the organization's communication style, dress code, and work schedule.

To move you along, more senior executives must know you, they must know your story, and they must have a relationship with you that allows them to stick their necks out for you, valuing the relationship enough to spend some of their own political capital on you. Often, this comes down to finding common ground. People with the power to see you are looking for areas of common interest, such as hobbies and shared values.

My early idea of success was firmly rooted in keeping my head down, keeping my mouth closed, letting my work speak for myself. I was often unwilling to socialize with my peers, thus not allowing them to get to know me, to see me. Nature abhors a vacuum. So it is in science, so it is in life! When you don't tell your story, others will fill in the empty space themselves, creating details based on nothing more than assumptions, preconceived notions, or outright biases.

It became more apparent to me, the higher I rose in my career, that people who were more open and willing to share themselves, advanced. Until my time in the ITSMF Executive Academy however, I hadn't integrated this knowledge into my work style. For someone like me, a self-styled "lurker" and lifelong doer, this one revolutionary fact changed my outlook. Now, I've shifted focus from what I can do to how I can build relationships and discovering what we can do together. I've found that I can reach out, open myself up to people, talk about my interests and experiences, and be vulnerable. An interesting reaction seems to keep coming up, something like, "You know, Shari, you're all right!" which sounds to me like, "Before I got to know you, I had my doubts about you!" These kinds of reactions weren't spurred by anything in particular that I had said or done, but what I was no longer not saying or doing: gone was my complete absence from the human conversations that connect us to others. Allowing people to learn about me and see my vulnerable side doesn't always feel natural, so I remind myself every day to stop and talk, ask questions, listen, engage.

Leadership is Not Managing

When given the opportunity to supervise a team, I excelled at coordinating and directing the work of others to achieve specific goals and objectives. I was good at establishing systems, processes, and structures to ensure that work was completed efficiently and effectively. I worked hard at planning, organizing, and controlling the work of others. I was a good manager; I am convinced of this. But leading and managing, while related, are not the same.

Leading involves inspiring and guiding individuals or teams to achieve a common vision or goal. A leader focuses on creating a shared sense of purpose, fostering motivation and commitment, and building relationships with followers.

As a manager, I relied on my authority, rather than using influence and persuasion to achieve results. Now, I keep focused on people and relationships. As a manager, a good manager, I often struggled with how to motivate my team to work as hard as I did. Now I understand that leaders don't motivate people. At least not the way that most of us think. Rather, successful leaders communicate a vision that people can buy into, a clear vision consistent with their own vision of themselves or where they want to go. Many team members are not necessarily moved by a goal, but by the *why* behind the goal; they want to connect to a leader's description of why the goal is important, how it fits into the leader's vision. At the heart of it, we're all self-interested individuals. We only buy into someone else's vision because it fits with our own vision, goals, and desires. A leader with a strong vision can articulate ideas and communicate their vision in a way that engages others, building a shared sense of purpose and direction, ensuring that everyone is working towards a common goal.

When I practiced creating a sense of excitement and enthusiasm among my team, helping them to see the value and importance of their work, I was much more successful as a leader. Having a clear and compelling picture of the future I wanted to create and the direction in which I wanted to take the organization, I was able to inspire and guide the team, helping them to see the bigger picture and their role in making it a reality.

There's an expression that goes, "You can work me long, but you can't work me hard." In other words, you can exercise some amount of authority over a worker, but you can't regulate their dedication or work ethic. Ultimately, I learned that only the employee can decide whether to use his or her skills and abilities, to give 100%, 100% of the time.

Leaders are Not Impermeable

Watching my father single-handedly raising two girls, I saw his strength and confidence. In our family, weakness wasn't tolerated. My vision of a successful leader was someone strong and commanding, so I molded myself into that vision. Yet while fortitude and leadership presence are important, I now see it's also important for leaders to be vulnerable because it allows them to connect with their team in a more authentic and human way.

When leaders are willing to share their own experiences, challenges, and emotions, they demonstrate that they are approachable and relatable, which helps build trust and rapport, and fosters a positive work environment. Being vulnerable lets leaders be seen as human and fallible, reducing the gap between leader and team, increasing the team's willingness to work with the leader, and fostering a sense of shared teamwork and collaboration.

I've also come to understand that vulnerability demonstrates courage and bravery. My willingness to take risks and admit to mistakes inspires others to do the same. I use this to create a culture of openness and honesty that can lead to more creative and innovative thinking. Acknowledging my own vulnerability has also helped me to better understand the perspectives and experiences of my team. My willingness to listen and be open lets me gain new insights and ideas. By more purposely connecting with my team, I've been able to better build trust, foster collaboration, and make informed decisions, making me a more effective and successful leader.

Being vulnerable is not being weak. I have learned to combine my own vulnerability with storytelling to shape a picture of myself for others. In order to follow someone, it's important to understand who they are, where they want to go, and why. Leaders who are adept at creating their own story are most successful; if you don't tell your own story, in an authentic and meaningful way, others will create a story for you.

Play the Game, or the Game Will Play You

Succeeding in corporate American, often described as a game, involves navigating complex rules and dynamic situations, making strategic decisions, and competing for resources and support.

Early in my career, I often said, "I just want to do my job; I'm not here to play games." But now I see that if you don't actively participate in the process or system in which you are involved, it will control and dictate the outcome for you. Or as one Executive Academy guest speaker put it, "If you don't play the game, the game will play you." Without taking charge of your own situation and making deliberate choices, you are at the mercy of external factors and forces. To succeed and achieve your goals, be an active player in the game and make strategic choices that will lead to desired outcomes.

Sharon L. Bradford

Shari Bradford is a Director, Lead Business Operations Associate at Nasdaq, Inc., where she works on the Technology & Operations Team in the Legal Risk & Regulatory Group. A 19-year veteran of Nasdaq, Ms. Bradford supports technology systems for the Office of General Counsel and the Investigations & Enforcement teams.

Leadership: It Only Takes All We've Got

by Tawanna Grant

One formal definition of executive leadership is the ability of those who manage or direct employees in an organization to influence and guide these individuals in fulfilling organizational goals, strategic planning development, and overall decision making. All true. And I'd add, an executive leader requires considerable intentionality to detail. As I found when I became an executive leader, we face multiple challenges, and must cultivate and grow many characteristics, including time-management, prioritization, strategic thinking, decision-making, problem-solving, and mature soft skills.

I've learned over the years to be flexible, to adapt and to do so gracefully. While I've worked in corporate America for decades, every scenario is different and what worked previously may not work again. Pivoting quickly can make the difference between a win and a loss. I've had to evaluate resource alignment for agility and capabilities to ensure alignment for the best outcome. Simply having the skills isn't enough to ensure success. Leadership requires being comfortable with being *uncomfortable*. Having enough courage, dexterity, and character to take a risk and hold your ground when needed. I've seen this happen when asked to pull in program dates or accelerate activity, and my experience may tell me it isn't good timing. That's when leaders make the decisions that are unpopular, but necessary. I encourage myself that I'm not starting from scratch, I'm starting from experience.

More Highs, More Eyes!

Every level of leadership requires some level of transformation—a continual process. Transitioning from middle-level management—being a doer and manager of tasks execution—to being an executive leader—a strategist and a solutionist, managing large scale efforts—requires the ability to leverage diverse opinions, build relationships, and evaluate decisions. Continuous development is the nourishment needed to remain sharp.

I've found that reading books on a variety of subjects enables well-rounded awareness. Expanding my network to include people who don't look like me or think like me gives me a foundation for innovation and creativity. What I listened to, I began to understand. Strategizing to win starts with understanding the game. *What are the rules of engagement? Who are the major players?* Knowing who to ask can save time and effort. I've learned to identify the decision makers and influencers quickly and to engage in relationships with some speed. The power to influence change toward a successful outcome has become my greatest accomplishment, usually because of knowing the difference between movement versus momentum and output versus outcome. You have to know *where* you're going to ensure you make it there *on time.*

Communication and maintaining open dialogue help to ensure everyone is in alignment. Goals are purposely measurable, achievable, and discussed continuously within the team. Teams gain momentum by understanding how choices influence the trajectory of achieving the goal. There's little to no benefit in having a great plan if it doesn't support and accomplish the goal. And we all know how great things may look on paper, but execution matters. Check and double check! I was once working on a project, ensuring all of my teams were on course but when we arrived at the big day, I had several stakeholders knee deep in the weeds. We had the train at the station, but none of the passengers were ready to board. This was a time when I definitely had to lean in and lead forward. I sharpened an innate ability to transition and refocus when necessary and to ensure engagement of others who are critical to the overall success. No woman is an island.

Leading others is complex and challenging, but well worth the effort. I tend to thrive in the areas of crafting solutions and managing effective change, which includes managing office politics and building strong relationships, gaining the appropriate resources that can assist in facilitating change, and navigating the complexity of human capital. It's important to be aware of transitions occurring within the organization and externally that are influencing outcomes. The best advice I've given is that the power of change is already within us. I choose how I show up.

I'm deliberate with decisions, recognizing that a "yes" to one action equates to a "no" to something else. I protect my personal brands to ensure that I share my narrative. I'm the source to knowing who I am, so why would I give that power away to anyone else? I've learned to ascend to the occasion. If the door is open, great! But sometimes you have to knock or even kick it down, create new bridges, build stronger foundations. I've planned for success while recognizing life is full of ups and downs, twists and turns, and I've learned just as much from what didn't go well as from successes. I'd had to improve, enhance, and evolve. Failure is a natural part of life, but as an executive leader we must make choices to "fail forward." Change is inevitable; we can prepare by looking for cues, reevaluating strategies, and pivoting when necessary, so that we thrive through change. Recently I led teams through a merger, opened new businesses, started on my doctorate degree, and began leading a new ministry, all in the same year. Another thing I've learned: the critical importance of prioritization and rest!

By participating in the ITSMF Executive Academy, I've gained a deeper respect and appreciation for executive leadership responsibility as a discipline, its complexity, and how much attention it requires. Being an executive leader demands a level of dexterity, resiliency, and commitment and many might find the responsibility and accountability to be daunting, difficult to obtain, and certainly tough to maintain. Yet, I find it an honor and privilege to achieve this level of leadership. We have what it takes to not only survive, but to thrive. It just takes all we've got!

Tawanna R. Grant

Tawanna Grant has worked in the financial services industry for over thirty years, currently serving as Senior Vice President, Software Engineering Director II at Truist, where she manages Treasury Technology Solutions-Onboarding and Data Delivery.

100% Lessons from Fortune 100 Leaders

by Jarel Hawkins

Executive leadership is the key to the success of any organization, particularly Fortune 100 companies that operate on a global scale and face complex challenges. Individuals in executive leadership positions are responsible for creating a vision and ensuring that it is communicated effectively to all stakeholders, including employees, shareholders, customers, and the broader community. They make critical decisions about resource allocation, talent development, risk management, and other key areas.

I know this now. But I didn't start out knowing it.

I am the middle child of childhood lessons; my parents were in corporate executive and educator roles, and they helped shape my passions. Growing up with an executive and an educator gave me the privilege of observing corporate culture, politics, and the value of teaching others new concepts. My childhood life lessons, combined with a passion for problem solving, helped develop my core values—transparency, accountability, creativity, and transformation. Now, I have notched over 20 years of multi-industry experience as a transformational change agent in the capacities of engineer, advisor, consultant, entrepreneur, and executive with multiple Fortune 100 and industry leading companies, including Altria, Pepsico, UPS, Truist, UBS, General Electric, Johnson & Johnson, FISERV, LVMH, Amazon, Home Depot, NIKE, BestBuy, and 3M.

Each company had a different culture and presented a unique set of challenges and opportunities. One of the most significant and common challenges I faced as a young Black executive was overcoming biases and stereotypes that at times didn't allow others to see me as an effective leader. Research has shown that Black men are often perceived as less competent, less intelligent, and less trustworthy than their white counterparts. This can make it difficult for skilled and qualified Black executives to be taken seriously.

Early in my career I saw an opportunity to prove what I thought were leadership abilities by working even harder and striving for excellence in the speed and quality of the work I did. I thought that by showing that I was capable and competent, I could break down the stereotypes and biases I thought were holding me back. I slowly but surely realized that doing things well just gets you more things to do! After a series of career changes, I concluded that a key challenge was lack of representation and role models of people of color in leadership positions; research bears this out across many industries, including technology and startups. This made it difficult for me to see myself as a leader until I realized the role sponsorship and mentorship played in the executive leadership journey. I then set out to listen and learn from great leaders and ask for support from them in my personal and professional evolution until I saw the CxO in myself.

I went on to successfully master the "two to three years up or out" career strategy and developed a leadership philosophy: ask, relate, educate, reiterate, accelerate, and innovate to build relationships with my teams rooted in trust, clarity, and transformation. This has helped me transform the challenge of underrepresentation into opportunity time and time again as a trailblazer and turnaround tactician.

The Skills to Scale or Fail

The *Harvard Business Review* reported that effective executive leadership involves three key elements: strategic clarity, disciplined execution, and cultural coherence.

The executive leadership team of a Fortune 100 company is responsible for setting the strategic direction of the organization and ensuring that it is executed effectively. This involves making critical decisions about resource allocation, prioritization, risk management, and talent development. Effective executive leaders must be able to balance short-term and long-term goals, as well as the needs of various stakeholders, including employees, shareholders, customers, and the broader community. At the same time, they must also model the values and culture of the organization and create a vision and a mission that inspires employees and aligns them around common goals. They lead by example, exhibiting integrity, transparency, and a commitment to excellence.

In a Fortune 100 company, executive leadership is characterized by a number of key qualities, beginning with building and managing teams. Leaders understand the importance of diversity and inclusion in team building, and work to create a culture of collaboration and innovation, ensuring that the right people are in the right roles, and that team members have the resources and support needed.

Fortune 100 leaders are focused on results; they set clear goals and metrics, hold themselves and their teams accountable, continually monitor and evaluate performance, and make necessary adjustments to ensure staying on track.

Finally, effective executive leaders in a Fortune 100 company are excellent communicators clearly able to articulate their vision, goals, and expectations to employees and stakeholders. Listening to and responding to feedback is also valued. These top leaders also understand the importance of transparency and honesty and are willing to acknowledge mistakes and take corrective action.

While executive leadership in a Fortune 100 company has many benefits, it also comes with challenges, the biggest being the pressure to perform. As a publicly traded entity, a Fortune 100 company is accountable to shareholders and under constant scrutiny from the media and other stakeholders. This puts pressure on executive leaders to deliver consistent financial results and to maintain the company's reputation.

The complexity of a Fortune 100 company is another challenge; leaders must navigate diverse business units and operations in multiple countries. This can make it difficult to stay on top of the many moving parts of the organization and to ensure all are aligned with the overall strategic vision.

Fortune 100 leaders are also tasked with balancing short-term and long-term priorities, delivering consistent financial results in the short term, and safeguarding the organization's long-term health and sustainability. This requires strategic planning and investment in research and development, talent development, and corporate social responsibility.

While the meaning of leadership can vary greatly depending on one's perspective and life experiences, executive leadership, in its simplest form, always involves the art of guiding and directing a company's strategic goals and initiatives through others. A successful executive leader sets a clear vision for the organization, drives a positive and engaging workplace culture, and empowers employees to achieve their full potential.

Jarel J. Hawkins

Jarel Hawkins is an emerging technology executive and advisor with 22 years of industry experience across supply chain, financial services, retail, manufacturing, real estate, healthcare, and technology.

Leading Deliberately Trough Sacrifices & Support

by Stephanie Kearney

"Sacrifice is the act of giving up something valued for the sake of something else regarded as more important or worthy."

- Amy Rees Anderson, *What Awesome Looks Like: How to Excel in Business and Life*

"There can be no progress, nor achievement, without sacrifice, and a man's worldly success will be in the measure that he sacrifices..."

- James Allen, *As a Man Thinketh*

In many teachings on executive leadership, the focus centers around the why and how of becoming an executive leader. In my experience, I've found two fundamental principles were paramount—sacrifice and support. Leaders must know how to make deliberate choices about the sacrifices they will make and the support they seek.

My first job after college was as an entry-level programmer in information technology. As a contributor, I had a planned yet flexible schedule, working an eight-hour shift between 6:30 am and 5:30 pm during weekdays, with on-call and early Sunday morning once-per-month rotations. I was responsible for only my own work product, achieving the goals explicitly assigned to me. My youth and simplistic life, the standard workday shift, schedule flexibility, and autonomy of my role made the sacrifices I had to make periodic and small, impacting only me.

For much of my IT management career, I continued to believe I could do it all and did do most of it, choosing to sacrifice things that affected only me. I rushed through the weekdays, constantly code-switching between "spouse-mother" and "work manager" roles based on my environment. Because my work schedule was predictable, with only advance-planned travel as needed, and since my spouse ran his own company and had flexibility, and my children attended school in addition to before-school, and after-school programs, I was frequently relieved from playing both roles simultaneously. While at work, I focused wholeheartedly on just being a "work manager." Likewise, I immersed myself in my "spouse-mother" role at home until spouse and children turned in for the evening.

Although weekdays were exhausting, the few hours in the late evenings reserved just for me-time gave me small pauses for relief. The weekends provided the ample time I needed to recoup, regroup, and reenergize before kicking off the cycle all over again.

With leadership however, the first change in thinking is understanding that being a leader is not a Monday through Friday nine-to-five job requiring only a few late evenings or weekends. Being a leader is a lifestyle, demanding more of a person's time, with the added pressure of people always watching what you do, how you do it, what you say/do not say, and how you say it. Leaders are like the E.F. Hutton of an organization: When leaders talk, people listen.

Leaders represent the company for those within and outside the company, including members of the local community(s) and other leaders in the industry. This always-on approach for a leader creates a blurred line between personal and professional life, making leadership a lifestyle that requires deliberately choosing what to sacrifice for self-preservation and continued success.

***

Here are a few examples of purposeful sacrifices I have made along my leadership journey—and why. These are examples of things I deliberately gave up because, for me, they were negotiable.

Sacrifice #1: Relinquishing the weight of my own voice in how others deliver the vision and strategy to articulate clearly and concisely what I believe.

Sitting in a meeting with my direct reports one afternoon, one manager confidently and passionately presented their three-year capability roadmap proposal that would enable our team to deliver our organization's strategy. Fifteen minutes into me listening with a keen ear, this person reached one bullet on the slide and rambled on for a few minutes. To better understand this point in the presentation, I asked, "When you say X do you mean Y?" The manager gave a definitive quick head nod and responded with a resounding yes. The rest of the presentation went on for another 15 minutes, thoroughly describing the outcomes and benefits, the total investment required with the positive return on investment, and one year payback period. I nodded and gave them a thumbs up to proceed to the following stages of our delivery process— the steering committee proposal review and approval.

In the pre-read this team member had sent out for the next meeting, I noticed something different about the roadmap. They had replaced a whole section with new capabilities. I immediately called them into a meeting to walk me through what had changed from the initial proposal and why. To my surprise, they had changed one of the capabilities to include what I had "told" them in our previous meeting. They elaborated that my "requested" change made via the one question I had asked altered all the downstream capabilities, which were dependencies.

First, I apologized for inadvertently driving the solution differently. To reframe our discussion, I started emphasizing why I believed in the strategy. Then I began asking how this new proposal would deliver on our strategy and align with the previous benefits and financial commitments. They concluded the conversation with a self-driven affirmation that they needed to update the capability section of the deck to reflect the initial capabilities listed and then send out a revised pre-read.

Sacrifice #2: Relinquishing being aloof and private and instead embracing vulnerability and taking a human-first approach.

"We all put on our pants the same way, one leg at a time."—Anonymous

It is easier for people to follow likable and relatable leaders. I managed my full team meetings with a strategic agenda, spending the first five minutes providing an overview of the agenda, jumping right into a deep dive on each time-boxed topic for the next 45 minutes, and ensuring that the meeting ended precisely five minutes before the scheduled meeting end time.

I knew everyone was busy, so I implemented this meeting approach to ensure the most productive use of everyone's time. Unintentionally, this resulted in my team giving me low ratings on an internal survey question, "I believe my manager cares about me."

As a leader, I learned that we must take a human-first approach, building time into meetings to tell personal stories that demonstrate our vulnerability and to check in with each member of our teams.

Sacrifice #3: Letting go of my discomfort of returning to the office so I could be physically present before, during, and after critical conversations.

Mid-2022, workplaces started opening back up after the 2020 pandemic. Like many others, my company had developed a hybrid work approach, making coming into the office optional. I went into the office only twice that year, once for a leadership meeting held during the middle of the week and once on the weekend to pick up a few items from my office.

Although well intended, hybrid meetings quickly create unintended divisive and ineffective collaboration with those of us participating via video, vying to get a word in. Others attending in person tended to get caught up in the topic, clattering amongst themselves, drowning out those on the call. Periodically, I needed clarification to understand the context of conversations that would occur in person before and after the session ended, often affecting critical discussions.

I had to acknowledge two things. First, I needed to find a way to continue fostering fundamental human connection with my peers. Second, I had to be an active and heard participant in the room going forward. Therefore, I had to stop considering being physically present for these meetings and before/after-hour social events as optional and attend in person.

If the head of our leadership team and members could travel from states away to participate live in these periodic meetings and events, my unwillingness to resume my 30-minute commute to attend in person seemed ridiculous.

<u>Sacrifice #4:</u> Replacing mindless time with time for maximized individual learning and growth.

Leaders must stay updated, relentlessly scheduling time to be more informed about themselves, the business, the industry, and world events. To unwind after a busy day/week, I often spent countless hours binge-watching or rewatching favorite TV shows and movies. I reserved my weekend evening for power reading bestsellers on the *New York Times* list. Although relaxing, this time collectively amounted to twenty or more hours a week indulging in nothing that would better my understanding of myself or the world around me.

I realized I could be more informed if I sacrificed this time spent escaping into a mindless zone. I shifted to having my TV tuned to news channels, rotating them throughout the day and evening to remain intentionally informed about local, regional, national, and worldly happenings. Also, I committed to reading two self-improvement and leadership books a month and incorporating reflection via journaling to memorialize critical concepts and learnings.

Support

What follows are some ideas, mindset, and activities I am now unwilling to sacrifice because they are core to who I am, my values, and my beliefs. These things are non-negotiable to me.

At some point, I established a deep-rooted narrative that I must "do it all" for the world to see me as a self-sufficient, high-performing, and successful woman. Now I know that effective leaders know they risk continued success in their personal lives and careers if they lead with this narrative.

But *doing it all* is a destructive lie. Even trying to do most of it, we risk inadvertently sacrificing spiritual, mental, and physical well-being. Without all three of these intact, leaders cannot show up as their entire best selves and achieve their purpose. To achieve this, I have come to realize the importance of leveraging the support of others.

"Don't be afraid to ask questions. Don't be afraid to ask for help when you need it. I do that every day. Asking for help is not a sign of weakness, it's a sign of strength because it shows you have the courage to admit when you don't know something and that then allows you to learn something new."- Barack Obama.

Support #1: Purchasing available services to shift the burden, providing a hyper-focus on still achieving the intended outcomes.

I used to pride myself on being a provider who also did everything in the home. I cooked homemade meals for my family at least five days a week and spent early Saturday and Sunday mornings cleaning the house, pruning the flower beds, raking the yard. I reserved Sunday afternoons for perusing every aisle in the grocery store for the following week's meal plan.

But leaders must learn to invest in products and services to realize desired value and benefits, so I shifted to paying companies and individual who specialized in providing these value-added services, to realize the value and benefit of spending with my family and on self-improvement.

Support #2: Leaning on my cheerleaders for support.

In the past, when the chips were down and I felt overwhelmed and undeserving of my blessings, I would tackle the challenge alone. I would enter a litany of self-talk, telling myself things like, "God puts no more on us than we can bear," or "Put on your big girl pants," and that I should "button up and get it together."

Yet leaders know that in times of doubt or challenge they need the support of a trusted inner circle, people who will have their back and provide warm words of encouragement and assurance. It has taken time, but finally I know to not white-knuckle my way through and bear the weight of my burdens alone. I have become vulnerable to my trusted circle of friends and selected family members and shifted from a squad of one to having several cheerleaders. They are my confidants who reground me on who I am and what I can do, what I should do, and reinforce that I am never alone.

***

Once I identified the areas of sacrifice and support in my life, I was able to better navigate my leadership journey and welcome growth, evolutions in my career, and personal advancements as they came my way. Through sacrifice and support, I am becoming the leader I want to be.

Stephanie L. Kearney

Stephanie Kearney is IT director of portfolio and program management at Hallmark Cards, Inc. She is responsible for the technology and digital workplace services portfolio and ensuring the timely, efficient execution of its associated projects and partnerships.

A Running Story: From One Track to Another

by Paul Lewis

BANG! After the gun fired, I held my breath for the next nine seconds in suspense to see what would happen next. The year was 1988, the place was Seoul, Korea and the event was the 100-meter dash final at the Olympics. The epic race between Canadian Ben Johnson and American Carl Lewis (no relation) ended with Johnson apparently winning the gold medal—until Johnson was stripped of his gold medal just three days later for testing positive for steroids, and Lewis prevailed.

At ten years-old, I knew how I wanted to lead—from the front of the pack on the track. From that point on, nothing else mattered; track and field became my Earth, sun, moon, and the stars. I developed an incredible work ethic centered around what I had witnessed, the eventual victor Carl Lewis working hard and overcoming incredible odds to achieve the sport's highest accolade, the Olympic Gold medal. Over the next fifteen years I worked harder than everyone else around me. In the early years, I developed my craft by racing anyone at any time—racing my brother from the car to the porch every time we returned home from anywhere, racing teens from my church in the parking lot or the field after church on Sunday. Whatever it took, I was determined to be the best.

My efforts and determination in track and field were highlighted with two-time Division 1 All American Honors, becoming James Madison University's record holder for the 200- and 400-meter dash, and qualifying for the U.S. Olympic trials. Teammates looked up to me as a leader, recruits who visited the University wanted to be a part of the program based on my success.

After a disappointing loss in the 2000 Olympic Trials, and failing to qualify once again four years later, it was time to utilize my degree in Computer Information Systems and get serious about my career. With the blueprint already in hand, I thought all I needed to do was transfer my work ethic and the success I achieved in my previous athletic life to the corporate environment.

I embarked on a journey to make up for lost time by reading every book I could and attaining technical certifications. While I was an entry-level support engineer at the Pentagon, my mentor laid out the educational pathway to close the knowledge gap between me and more seasoned engineers. I was focused on obtaining a Microsoft Certified Systems Engineer 2003 (MCSE) certification and paid in advance for all seven tests, taking one test per month. Within a year I became the Deputy Lead of Operations and things were looking up. My work ethic led to an opportunity to transition to Information Security at the Pentagon; I was on a roll. Sticking to the script and following the same blueprint, CompTIA Security+, ISC2 Certified Information Systems Security Professional (CISSP), and SANS certifications were next up. Positions and money came quickly; however, my frustration began to grow because my hard work never got me past a mid-level team lead role.

Then something strange happened. My next two moves to a director role came from the result of relationships that I built, not all that "hard work." These positions came with more responsibility, paid more, and I barely interviewed for them. Initially, I felt insecure in these roles because they weren't roles that I "worked my way into."

An enlightening read of Harvey J. Coleman's book *Empowering Yourself: The Organizational Game Revealed,* confirmed for me that for all of my hard work and efforts, all of those years, I was nothing more than a "resource" to the people running in the real race. I wasn't on the right path or establishing the right relationships to be mentored and eventually sponsored by the real players in the game. This shattered my thirty-year blueprint for success, which was: work hard, the right people will notice, and eventually I would move up. Harvey highlights that out of performance, image, and exposure (P.I.E.), the weight relative to the impact on promotion is, performance is only 10%, while image counts for 30%, and exposure, a whopping 60%. My image and exposure needed work.

I always heard about the "other race," but I wasn't sure exactly how to get in it. The executive leadership race requires more than working harder and running faster than the person next to you; this race takes place on many levels. Realizing exactly where I was and what it takes to move to the next level was eye opening.

Back to the Future

If I were Marty McFly and I could hop in the DeLorean to go back in time to rewrite my blueprint, I would have told my younger self these three things: Develop a career plan. Find a mentor (ask for help). Work on my image.

Working hard and hitting your organization's key driver goals alone is not a career plan. *Where do you want to be in five years? Where do you want to be next year? What do you need to do today to prepare for that? Does that destination require additional education? Does it require that you move into other roles in your organization to broaden your knowledge, skills, and ability in the enterprise? Have you looked at the job description for the role that you want in five years? How much of that experience do you have today?*

Once the career plan is in place, the plan needs to be communicated with your manager. In addition to meeting the organizational goals, what stretch assignments can you work on to help round out the experience that you are seeking to meet your greater career goals?

After developing a career plan, identify mentors that could help you navigate to get to the destination, and communicate your plan. I now reflect on the missed opportunities from not establishing relationships with key government officials and consulting executives that had literal open office doors that I walked past every day, people who would have been great resources. These connections could have led to mentorship, sponsorship, or consideration for business opportunities in consulting. Earlier in my career, I really needed a mentor. There were senior engineers and analysts that I developed working relationships with that I could lean on for technical advice when I found myself in a jam, but I did not seek to establish a relationship with "the guy I wanted to be when I grow up."

Early in my career in government I vividly recall working for a division chief who oozed power from his pores as soon as he walked into the room. Our weekly meetings always took place in a conference room with twenty seats. He sat at the head of the table, and there were always two empty seats on either side of him; no one dared to sit any closer for fear that a spark from the flames that he sometimes spewed would touch them. His reaction and response to answers to his questions that were not satisfactory made everyone uncomfortable. Instead of being intimidated by him, I should have asked him for mentorship.

When I worked in financial services, the Chief Information Officer (CIO) hired a superstar to lead engineering, someone everyone knew would eventually become the next Chief Information Security Officer (CISO). He and I were both attending a vendor sponsored event and had an opportunity to talk for an extended period. When the event ended, we sat outside talking for another 45 minutes, and he shared some career advice; it felt as if I'd made a new friend. He extended an invitation to connect with him in future, saying, "Any time you want to talk about anything just reach out." Needless to say, today he is a CISO and doing very well.

At times, I felt like the minister in the "two boats and a helicopter" parable. A minister refuses several offers of help in the face of approaching floodwaters, each time telling the would-be rescuers that God will save him. After turning down the last attempt, he drowns in the flood. After his death, the minister meets God and asks why He did not intervene. God responds, "I sent three rescuers to your aid. Why didn't *you* accept the help?"

Soon after entering the ITSMF academy, I asked myself why I am not further along in my career. Looking back, I see that God sent me two boats and a helicopter and I didn't recognize that they were there to help me.

During the ITSMF Executive Academy, Harvey Coleman dove deep on image, covering many aspects including physical appearance, speech, tone, and attitude. The image adjustment I needed to change the most was my image of myself. I never saw myself as an executive leader or felt as if I belonged in the circle with them. In past conversations with people senior to me, I would usually ask them a version of the question, "How did you get to where you are today?" When I heard their story, I'd be inspired, but many times afterwards I would focus on the one or two things that they had that I didn't and discredit myself. *Oh, I didn't graduate from an Ivy League school. I wasn't an officer in the military. I didn't come from the top investment banking firm.* I needed to develop a stronger self-image. Having a positive internal image is the first step to creating the image that you want others to see. Now, when I wake up and look in the mirror, before I engage anyone else, I see a leader, not someone less than.

I recently had a conversation with my CIO who is new to my corporation. I asked her about the things that she did in her past to help prepare her for her current role. She shared a bit about her background and everything she did to prepare—the various roles she held in previous enterprises, an MBA, all fit into her plan to become a CIO. I also asked how she was able to transition from another industry sector and make critical strategic decisions so early on. She told me that she has several peers in the industry that she regularly reaches out to, to see how they handle similar challenging situations. Hearing how she prepared for the CIO role with a career plan, how she continues to ask for help, was reassuring that I am on the right path.

All the puzzle pieces have always been on the table for me. The difference is that now I have a clear picture of what it looks like when they are all put together. I'm now on the right track.

Paul Lewis

Paul Lewis is Director, Information Security at BD. His cybersecurity experience includes roles at Ernst Young, BNY Mellon, and First Data Corporation.

Wild Dogs and the C-suite Choice

by Janene M Worthington

In October 2020, when the pandemic was raging and I was stuck in the house, I watched a NOVA documentary on PBS that unexpectedly opened my eyes: I had been winning my career *on a small scale*. I realized the playing field within my professional landscape is so much larger and the rewards are so much greater, if only I could unlearn some of my formative lessons that were getting in the way of bigger wins and keeping me at good instead of *great.*

The documentary, *Nature's Fear Factor*, was about the role of fear in the animal kingdom. Scientists were reintroducing predators, native African wild dogs, to Gorongosa National Park in Mozambique, Africa. In the absence of fear from the threat of predators, wildlife populations there had thrived and grown, unchecked. In the absence of the behavior-changing fear that predators and the threat of danger produced, populations had stretched to new territories beyond expected safe zones.

In the animal kingdom, fear is learned and can be passed from generation to generation. It is pervasive and instinctive, meant to protect and ensure survival. Once established, the landscape of fear keeps prey in check, even without the predator participating. Fear can shape the behavior of the many by impacting only a few.

This documentary about how fear changes behavior struck me and stayed with me. I spent time considering parallels between a landscape of fear for animals and a landscape of fear for humans. I thought about the impact of fear on humans, particularly myself. I wondered what fears were keeping me in check personally and professionally. What fears were holding me back from an aspiration to ascend to executive leadership and C-suite Level? What "wild dogs" of self-limiting beliefs and behaviors may be stealing this choice away without my even noticing?

I started a journey of identifying my own "wild dogs."

I needed to explore with clarity and vulnerability the lessons that I have internalized for my protection and survival "in the wild"—lessons imparted during formative years, repeated so many times or delivered so convincingly that they became instinct. These are the lessons that I have rarely questioned and had remained with me.

To find my "wild dogs" I had to start with knowing myself. Life experiences differ from person to person. I needed to objectively recognize, assess, and unlearn a few limiting beliefs to change my landscape of fear to a landscape without limits.

Wild Dog #1 | "What happens in this house, stays in this house!"

When you are raised with the mandate that what happens in this house, stays in this house, and you have an introverted personality, you learn to keep your family's personal lives to yourself.

When I was around the age of nine, someone broke into our house and stole my uncle's "component set" (have I dated myself?) and many of his records and albums. My mom and grandma said it happened because my uncle "told everybody everything." I internalized this story and determined that bad things happen when you don't keep your personal life private or you invite too many people into your space, so they know how and where you live, and what you have or do not have.

That cultural norm taught me, from an early age, to be wary of people outside my trusted circle of family and friends. For years, I was driven by this instinct personally and professionally. I kept my circle of friends small. I did not allow many people to know too much of my story, and I did not get to know them personally for fear that I might be asked to reciprocate.

Leadership Limitations of Wild Dog # 1: By restricting access to myself and turning down offers to socialize with diverse co-workers where I might have to talk about myself, my family, and background, I devalued my story and missed opportunities to learn about others. Executive leadership requires sponsorship. Sponsorship requires familiarity. To achieve familiarity, you must foster relationships with diverse leaders to discover commonality of interest and experiences.

***

Wild Dog #2 | "If you can't get it from us, you don't need it!"

When I was growing up, I learned that help comes from family and no one else. Help not given by your family was "charity" and indicated that you couldn't take care of yourself. Taking charity from neighbors in any form was unacceptable. Eating at a neighbor's house was the worst offense. Accepting a meal, if discovered, would earn you a very stern rebuke: "If you were hungry, you should have come home."

It is here that I learned to limit requests for help. Requests for help could result in pity or ridicule. I didn't want to feel indebted to people. I learned to rely on myself, to figure it out.

Leadership Limitations of Wild Dog # 2: Operating based on advice from family members who have not worked in your role or understand the rules of your professional ecosystem may not be the best choice. Everyone needs help. No one accomplishes anything on their own. Asking for help is not an admission of inadequacy; it is a tribute to people who want to assist. Refusing to recognize and admit when help is needed in small and big efforts burns up energy unnecessarily and restricts your capacity to think more expansively. Executive leaders all eventually understand that the only way to deliver their vision is with the help of other people. Motivating others to do more requires vulnerability and an earnest request for help.

***

Wild Dog #3 | "Has everyone eaten?"

My family couldn't always sit down to eat a meal at the same time together. So, before helping ourselves to food, we were conditioned to ask if everyone else had already eaten. If the answer was no, you knew to leave something for those who would eat later. There were no seconds until everyone had firsts. We learned to consider others. You ate less than maybe you really wanted, to ensure the needs of those who would eat later could be met.

Even though it may not have been intended, I routinely considered the needs of others first and deprioritized what I may have wanted or needed. I got the message that putting myself first was selfish and less than humble, and selfishness is always unacceptable.

Leadership Limitations of Wild Dog # 3. Attempting to convince anyone that I should deprioritize my wants and needs for professional development in favor of others is silly. Thinking about my own achievement as selfish is a smoke screen for self-doubt and protection against the possibility of failure. Effective executive leaders can and do embody confidence in their strengths and abilities coupled with humility to recognize that they have blind spots. Leaders doggedly advocate for their vision of the future and selfishly want to see that vision realized.

Overcoming Wild Dogs and Owning my Choice

By facing down my own "wild dogs," I now recognize that executive leadership is a choice, and completely within my power to attain. It has been necessary to choose a different mindset by unlearning behaviors that do not serve my purpose professionally and are not relevant in professional environments. I will need to continue setting aside certain social and cultural norms that I have always accepted and allowed to impose limits on my possibilities. Examining my stories, my wild dogs, exposes the myths that people with limited world views told me about how the world works and my place in it. They had good intentions, but I let their protections extend too far.

Ascending to the C-suite will require deliberate practice and effort. I will need to find clarity of purpose and vulnerability to share my leadership goals with others. To become a better leader, I will need to humble myself to ask for feedback and suggestions for improvement. The feedback may leave me feeling exposed or temporarily undermine my confidence, but I will reap the benefits when I listen with gratitude. My wild dogs taught me that I can't do it alone. I need to acknowledge differences and seek similarities with people who, at face value, and through conscious and unconscious bias (we all have it) seem completely different from me. The help I need to be an executive leader will result in connections and relationships that I would not normally seek and had not imagined possible.

The skills to be recognized for executive leadership may first be based on credibility that I gained through proficiency to deliver results. However, executive leaders require a broader skill set. Broadening and deepening my knowledge across various disciplines will influence my perspective and vision. The role of storytelling will also be critical to drive connections and motivate others to willingly do the work to deliver my vision. My ability to create connections by sharing my personal story, build and sustain relationships, and influence outcomes through others will be key to my success.

The requirements for executive leadership are within my control. At times, I admit, this assertion is not easy to believe. It feels like forces set in motion hundreds of years ago persist today for the sole purpose of keeping me from achieving abundant success. The "wild dogs" are always lurking to make me doubt that I am in control. It's easy to limit myself if I get drawn into someone else's version of truth, especially their definitions of the right schools, right neighborhoods, right connections, or right hobbies and interests.

If I am honest, I did not consider executive leadership in my early career because of assumed deficiencies I thought would handicap and disqualify me from consideration. I assumed executive leadership to be outside the realm of possibility, although no person other than myself ever told me that I was not or could not become qualified. I alone disqualified myself based on fears, conditioning, and unfounded assumptions; instead, I felt lucky to be given the opportunities that came my way.

I now know that this thinking is self-limiting and foolish. The negative stories that we tell ourselves stop any forward momentum before we have a chance to find out how great we can actually be. In *The Power of Choice: Embracing Efficacy to Drive Your Career,* Michael C. Hyter writes, "The power of choice is not about becoming successful in spite of your differences or even because of your differences. It's about expanding the definition of who you are and who you can become. It's about expanding your sense of possibilities for yourself."

As I look forward, I will take back my choice grounded in the real truth of the leader I am capable of being. I recognize new truths now.

I know now that I have earned every professional opportunity. If there was luck, you can believe that I met it at the crossroad of preparation. We have everything we need. There is nothing outside of our control that is obstructing our path. Choosing our mindset and being aware of the limitations we impose on ourselves is key.

The perception of risk, of fear, can keep us from exploring new territory, allowing ourselves to be satisfied with good enough when great is achievable. We may choose not to trust a situation based solely on the long-past experiences of others. The animal kingdom teaches us that even nutrition will be sacrificed in exchange for safety. Animals will eat less well in order to ensure their survival.

Which opportunities are being sacrificed? Left on the table because our perception of risk in the professional landscape has unfounded (having no basis in fact) or untested (not proved by trial or experience) limits imposed by learned behavior and/or past social or cultural norms? What are the wild dogs of your own personal past?

Janene W. Worthington

Janene M Worthington is Director of Data Engineering with Capital One Commercial & Small Business Banking Technology, committed to building high performing, inclusive teams.

Appendix

ITSMF | INFORMATION TECHNOLOGY SENIOR MANAGEMENT FORUM

ABOUT ITSMF
Mission, Vision, History, and Leadership

OUR MISSION
ITSMF increases the representation of Black professionals at senior levels in technology, to impact organizational innovation and growth. We do this by developing and nurturing these dynamic leaders through enrichment of the mind, body, and soul.

OUR VISION
To facilitate the creation of **barrier-free** opportunities for Black technology **influencers, innovators,** and **leaders.**

OUR HISTORY
The Information Technology Senior Management Forum (ITSMF), formed in 1996, began with a conversation between a few prominent technology executives who found that only 3 percent of information technology management roles were held by Black professionals. Studies suggested that not enough Black professionals were positioned for senior-level technology positions. This discussion sparked those executives to action as they were compelled to reach, teach, and cultivate those aspiring technology professionals.

Under the leadership of Carl Williams, the 501c (3) organization was founded in Chicago, Illinois and began the work of recruiting those Black professionals who ranked among the Who's Who in technology, and those companies and individuals looking to support the mission of the organization. ITSMF, celebrating its 20th year, remains the only national organization dedicated exclusively to cultivating executive talent among Black technology professionals.

Our Leadership
Board Of Directors

Michael Buckham-White
Chairperson

Managing Director,
Information Technology
Deloitte

Ibrahim Jackson
Vice Chairperson

Founder & CEO
Ubiquitous Preferred Services, Inc

Rona Bunn
Corporate Secretary and By-Laws
Chairperson

CIO
National Assn of Corporate Directors

D. Lena Hickman-Miott
Board Member

Chief Information Officer and Chief
Operating Officer
Miottech, LLC

Fern Johnson
Board Member

CTO Vice President
PepsiCo, Inc.

Theresa Wilson
Board Member at Large

(ret) Executive Vice President
Wells Fargo

Marilyn Crouther
Board Member

CEO and Principal
Crouther Consulting, LLC

Art Hopkins
Board Member

Global Technology Officers Practice Co-leader
Russell Reynolds Associates

Don Neault
Board Member

CEO, Certified Leadership and Executive Coach, and Corporate Advisor

Ehrika Gladden
Board Member at Large

Sorenson

Dane Bamburry
Finance Chairperson

Senior Director, Enterprise Architecture
Cox Enterprises, Inc

Bennette Veals
Management Associates Representative

Manager, Global Business Services
The Procter & Gamble Company

Da'Shaun Joseph
Associate Member Vice Chairperson

Program Manager
Google

Johanna R. Jones
President & CEO

Carl Williams
Chair Emeritus
(deceased)

Executive Team

Johanna R. Jones
President & CEO

Cathy Butler
Executive Assistant

Xenon Walcott
VP, Operations

Elaine Norman
Vice President of Member Services
and External Affairs

Michael Whitley
VP, Development and Advancement

Robert D. Scott
Vice President and Dean of the Global Institute for Professional Development (GIPD)

Dr. Reginald H. Turner
Management Academy Director

Dr. James Rodgers
Executive Academy Director

Kailei Carr
EMERGE Academy Director Emeritus

Cheryl Bridges
Incoming EMERGE Academy Director

Denise Smith
IGNITE Academy Director

Sherrell Bailey
Director, Events

Richard Brown
Director, Events

Cedric Cobb
Director, Development

Dawn Jordan Jones
Brand and Communications Manager

Nigel Cudjoe
GIPD Administrative Program Manager

Valda Cowan
Program Coordinator College
Mentoring Program

Ernestine Key
Accounting and Development Specialist

Made in the USA
Monee, IL
04 November 2026

67385850R00105